ATLAS
OF POINT CONTACT SPECTRA
OF ELECTRON-PHONON
INTERACTIONS IN METALS

ATLAS OF POINT CONTACT SPECTRA OF ELECTRON-PHONON INTERACTIONS IN METALS

by

A. V. Khotkevich and I. K. Yanson

*B. Verkin Institute
for
Low Temperature Physics and Engineering*

translated by

Randal C. Reinertson

University of Maine at Orono

KLUWER ADEMIC PUBLISHERS
Boston / Dordrecht / London

Distributors for North America:
Kluwer Academic Publishers
101 Philip Drive
Assinippi Park
Norwell, Massachusetts 02061 USA

Distributors for all other countries:
Kluwer Academic Publishers Group
Distribution Centre
Post Office Box 322
3300 AH Dordrecht, THE NETHERLANDS

Library of Congress Cataloging-in-Publication Data

A C.I.P. Catalogue record for this book is available
from the Library of Congress.

Copyright © 1995 by Kluwer Academic Publishers

All rights reserved. No part of this publication may be reproduced, stored in a retrieval system or transmitted in any form or by any means, mechanical, photo-copying, recording, or otherwise, without the prior written permission of the publisher, Kluwer Academic Publishers, 101 Philip Drive, Assinippi Park, Norwell, Massachusetts 02061

Printed on acid-free paper.

Printed in the United States of America

CONTENTS

Preface vii
Abbreviations and Basic Symbols x

1. THE METHOD OF POINT CONTACT SPECTROSCOPY 1

1.1 Models of Point Contacts 1
1.2 Regimes of Electron Transport Through a Constriction 2
1.3 Resistance at Zero Bias 5
1.4 Nonequilibrium Electron Distribution Functions 8
1.5 Principal Theoretical Relationships 10
1.6 The Electron-phonon Interaction Point Contact Functions 12
1.7 Heterocontacts 18
1.8 Two-phonon Processes 23
1.9 Comparison of Electron-phonon Interaction Point Contact Functions
    to Related Functions 24
1.10 Background 27
1.11 Point Contact Spectroscopy of Non-phonon Excitations 29
1.12 Methods of Forming Point Contacts 29
1.13 Quality Criteria 31
1.14 Modulation Methods for Measuring Derivatives of the
     Current-voltage Characteristics 34
1.15 Modulation Broadening of Spectral Lines 36
1.16 Block Diagram of a Spectrometer 37
1.17 The Procedure for Reconstruction of the Electron-phonon Interaction
     Point Contact Function From Measured Characteristics 39
1.18 Pseudopotential Calculations of the Electron-phonon Interaction
     Point Contact Functions 41
1.19 Methods for Determining the Electron-phonon Interaction Functions
     and the Phonon Density of States 42

2. POINT CONTACT SPECTRA, ELECTRON-PHONON INTERACTION
   FUNCTIONS, AND THE PHONON DENSITY OF STATES IN
   METALS 45

2.1 Lithium 45
2.2 Sodium 48
2.3 Potassium 51
2.4 Copper 54
2.5 Silver 60
2.6 Gold 62
2.7 Beryllium 64

2.8 Magnesium	66
2.9 Zinc	68
2.10 Cadmium	72
2.11 Aluminum	74
2.12 Gallium	77
2.13 Indium	79
2.14 Thallium	83
2.15 Tin	85
2.16 Lead	88
2.17 Vanadium	91
2.18 Niobium	94
2.19 Tantalum	97
2.20 Molybdenum	99
2.21 Tungsten	102
2.22 Technetium	104
2.23 Rhenium	105
2.24 Iron	108
2.25 Cobalt	109
2.26 Nickel	111
2.27 Palladium	113
2.28 Osmium	114
2.29 Gadolinium	116
2.30 Terbium	117
2.31 Holmium	118
References	120
Subject index	150

Preface

The characteristics of electrical contacts have long attracted the attention of researchers since these contacts are used in every electrical and electronic device. Earlier studies generally considered electrical contacts of large dimensions, having regions of current concentration with diameters substantially larger than the characteristic dimensions of the material: the interatomic distance, the mean free path for electrons, the coherence length in the superconducting state, etc. [110].

The development of microelectronics presented to scientists and engineers the task of studying the characteristics of electrical contacts with ultra-small dimensions. Characteristics of point contacts such as mechanical stability under continuous current loads, the magnitudes of electrical fluctuations, inherent sensitivity in radio devices and nonlinear characteristics in connection with electromagnetic radiation can not be understood and altered in the required way without knowledge of the physical processes occurring in contacts.

Until recently it was thought that the electrical conductivity of contacts with direct conductance (without tunneling or semiconducting barriers) obeyed Ohm's law. Nonlinearities of the current-voltage characteristics were explained by joule heating of the metal in the region of the contact. However, studies of the current-voltage characteristics of metallic point contacts at low (liquid helium) temperatures [142] showed that heating effects were negligible in many cases and the nonlinear characteristics under these conditions were observed to take the form of the energy dependent probability of inelastic electron scattering, induced by various mechanisms. One important mechanism is the scattering of electrons on vibrations of the metal lattice, i.e. the electron-phonon interaction, which generates all of the most important fundamental and practical properties of metals (superconductivity, for example).

Study of the nonlinearities in the current-voltage characteristics of point contacts at low temperatures has developed into an independent branch of the physics of metals, namely the method of point contact spectroscopy. The possibility of successful application to clean metals (and also to many alloys and compounds) and the relative technical simplicity of the method make it practically indispensable for determining the characteristics of the electron-phonon interaction, i.e. the spectral functions. These functions were previously well known only for a few superconducting metals. Results obtained by point contact spectroscopy can be utilized for calculations of the thermodynamic and kinetic characteristics of metallic systems (electrical conductivity, thermoconductivity, thermal expansion, etc.), as well as for verifying microscopic theories of metals and useful corrections to their input parameters.

Since in many cases the electron-phonon interaction function differs from the phonon density of states only by a multiplication factor with a negligible frequency dependence, point contact spectroscopy also permits study of the phonon density of states. Direct, non-model determination of characteristic properties using neutron scattering are exceptionally laborious and often ambiguous due to low precision and resolution. Reconstruction of the density of states through the phonon dispersion curve, measured by neutron scattering to high precision, requires the use of approximate theoretical models describing the dynamic interaction of atoms in the crystal lattice. In addition, large single crystals, which are not always available, are necessary for neutron studies while point contact experiments are practical with specimens having dimensions of fractions of a millimeter. For such specimens point contact spectroscopy presents a fast, simple and inexpensive (without the use of a nuclear reactor) method for investigating the phonon spectrum of metals.

In this book, general and systematized scientific information of the electron-phonon interactions in clean metals, selected as the most reliable data, is presented in a convenient form for comparison to related characteristics of the electron-phonon interaction obtained by other methods and the phonon spectra known for some metals.

This reference book consists of two parts. The first part is dedicated to a general description of the method of point contact spectroscopy and a presentation of the principal relations of the theory. The technology of forming point contacts and methods for measuring point contact spectra (dependences of the second derivative of the current-voltage characteristics on the voltage applied to the point contact) are briefly described and the relations between measured variables are indicated. The criteria for selecting point contacts according to their characteristics are set forth, allowing establishment of the degree to which real contacts conform to their theoretical models. The procedure for reconstructing the electron-phonon interaction point contact function from measurements of the point contact spectra is described.

In the second part, point contact spectra of the electron-phonon interaction in metals and the fundamental point contact functions obtained from them are presented. Graphs of related functions of the electron-phonon interaction (thermodynamic and transport) and the phonon density of states are inserted for comparison. Data is organized by material, with a separate section dedicated to each metal. In the first illustration of each section the un-retouched point contact spectrum of the electron-phonon interaction for the given metal is shown on a large scale with a coordinate grid for determination of the positions of spectral features relative to the intensities of different maxima and the background level. In supplementary figures the point contact spectra, measured to large voltages (containing, as a rule, the two-phonon maxima) and/or point contact spectra obtained by other authors using different techniques and spectra with different relative intensities of the maxima are shown. In addition to the figures, the characteristics of a point contact and conditions of measurement are listed. The point contact function is presented in the form of a graph and a table. In selecting graphs of functions related to the electron-phonon interaction (if these are known for the given metal) and the phonon density of states, preference was given to experimentally obtained data.

The reference list contains other literature on point contact spectroscopy as well as the majority of original journal articles devoted to determination of the EPI functions and phonon density of states in clean metals. For convenience, a literature list is placed after each section heading showing reference numbers under the headings: point contact spectroscopy, electron-phonon interaction functions, and phonon density of states. Numbers of references dealing with the theory of point contact spectroscopy, point contact spectroscopy of alloys, and point contact spectroscopy of semiconductors are contained in the literature list located at the end of the first part of the book.

The authors are grateful to academician B. I. Verkin and to their colleagues who assisted in producing this book.

The authors

Abbreviations and Basic Symbols

FCC	- face-centered cubic
hcp	- hexagonal close-packed
BCC	- body-centered cubic
FS	- Fermi surface
EPI	- electron-phonon interaction
a	- point contact radius
a, b, c	- lattice constants
B	- magnetic induction
$B(eV)$	- point contact background function
d	- point contact diameter
e	- charge of the electron
f	- modulation signal frequency
$F(\omega)$	- phonon density of states
$g(\omega), g_{tr}(\omega), g_{pc}(\omega)$	- thermodynamic, transport and point contact EPI functions
$G(\omega)$	- unnormalized EPI point contact function
i	- modulation current
I	- direct current
$j = 1, 2$	- number of the contact metal
k	- Boltzmann's constant
K	- point contact form factor
l	- mean free path of an electron
$l_i, l_\varepsilon, \Lambda_\varepsilon$	- momentum, energy and diffusion free paths for an electron
L	- length of a contact (channel model)
m	- mass of a free electron
$N(\varepsilon)$	- electron density of states
p	- quasi-momentum of an electron
p_F	- Fermi momentum
P	- pressure
q	- quasi-momentum of a phonon
Q	- reciprocal lattice vector
R	- differential resistance of a point contact
R_o	- point contact resistance at $V = 0$
s	- number of the phonon spectrum branch
S	- contact area
S_F	- area of the Fermi surface
T	- temperature
v	- electron velocity
v_F	- Fermi velocity

V	- voltage
V_1	- r. m. s. voltage of the first harmonic of the modulation signal
$V_{1,0}$	- r. m. s. voltage of the first harmonic of the modulation signal at $V = 0$
$V_2, \tilde{V}_2, S_2$	- r. m. s. values of the voltage of the second harmonic of the modulation signal, proportional to the second derivatives of the current-voltage characteristics d^2V/dI^2, d^2I/dV^2, and $d(\ln R)/dV$
z	- coordinate parallel to the axis of the contact
γ	- point contact background parameter
ε_F	- Fermi energy
$\delta(x)$	- Dirac delta function
Θ	- electron scattering angle
$\Theta(x)$	- Heaviside theta function
$\lambda, \lambda_{tr}, \lambda_{pc}$	- EPI parameters, corresponding to the thermodynamic, transport and point contact EPI functions
ν	- ratio of the momentum mean free path of an electron to the characteristic dimension of a contact
ρ	- resistivity
σ	- conductivity
ω	- phonon angular frequency
ω_{max}	- maximum frequency of the phonon spectrum
Ω_{eff}	- effective volume of phonon generation
$\hbar$	- Planck's constant
$\langle \cdots \rangle$	- average over the Fermi surface

ATLAS
OF POINT CONTACT SPECTRA
OF ELECTRON-PHONON
INTERACTIONS IN METALS

1. THE METHOD OF POINT CONTACT SPECTROSCOPY

1.1 Models of Point Contacts

Two bulk metallic electrodes, contacting one another (contiguously) over a small area, constitute an electrical contact of small dimensions - a *point contact* (from here on, so that there is no confusion, referred to simply as a contact). When electric current is passed through such a system, it is concentrated in a narrow region (*systems with a concentration of current*), reaching densities of 10^{13} - 10^{14} A/m^2. The metal in this narrow region is not overheated due to the effective heat flow to the banks (the electrodes) of the contact, provided that the mean free path for energetic relaxation of electrons is larger than the characteristic (largest) dimension of the narrow region.

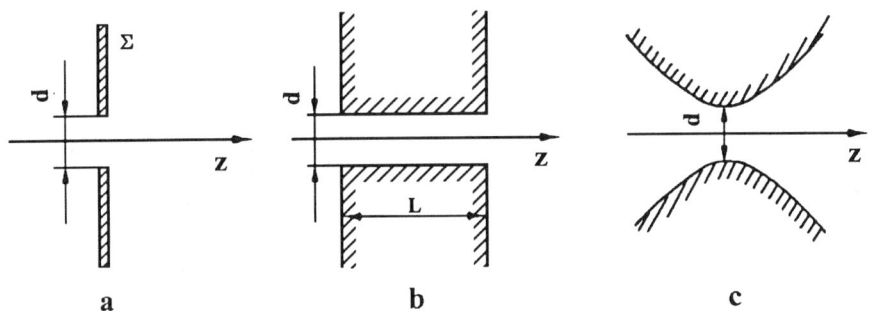

Figure 1
Point contact models: a - aperture   b- channel   c - hyperboloid of revolution

For real point contacts the following basic theoretical models are used.

1. *The model of an aperture* with an infinitely thin (but not transparent for electrons) flat partition Σ, dividing the two metallic half-spaces (Figure 1a) [125]. The aperture is of a basic form, i.e. a rectangle, circle, or ellipse. Accordingly, the characteristic dimensions of such contacts are the square root of the area of the rectangle, the diameter (or radius) of the circle, and the mean diameter of the ellipse.

2. *The model of a long channel* [68] (the length of the channel, L, is much larger than its diameter, d), filled with metal and connecting the bulk metallic banks (Figure 1b).

In intermediate cases, a point contact is modeled as a single-volume hyperboloid of revolution (Figure 1c), [67] forming a constriction with a minimum cross sectional area of $S = \pi d^2/4$ and effective length $L = \sigma_0 R_0 S$, where σ_0 is the electrical conductivity of the metal, dependent on the elastic scattering of electrons, and R_0 is the resistance of the contact at zero voltage. It is assumed that the elastic scattering length (mean free path) of the electrons l_i, is much less than the maximum of L or d.

If the elastic scatterers are nonuniformly distributed in the constriction region, there are two different idealized cases for a contact in the form of an orifice.

The *T-model* [321] [or Z-model (158)] corresponds to scattering concentrated near the plane of the contact (Figure 2a) and constituting a barrier with a transmission coefficient (for electrons) of $T \leq 1$. Nonequilibrium phonons, which are formed by energetic electrons in the narrow region, freely escape to the clean edges.

The model with full or partial reabsorption of nonequilibrium phonons in the narrow region [64] (Figure 2b) corresponds to elastic scattering on the periphery of the contact and the return of nonequilibrium phonons into the narrow region. These scatterers have no effect on the electron distribution in the contact, as long as they are located in a region where the deviation of electron subsystems from equilibrium is small.

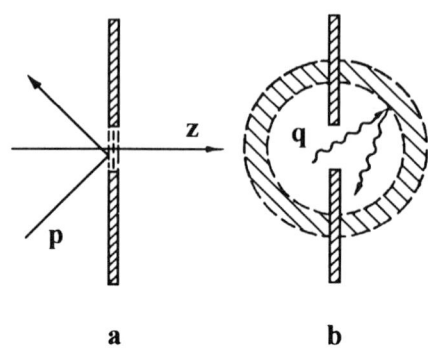

Figure 2
Models of contacts with elastic scattering: (**p** is the quasimomentum of an electron and **q** is the quasimomentum of a photon)
a - T-model
b - scattering concentrated on the contact periphery

1.2 Regimes of Electron Transmission Through a Constriction

In point contact spectroscopy, two possible asymptotic regimes can be distinguished for the electric current through a point contact.

The ballistic regime [338] (clean limit) corresponds to the inequality
$$l_i, l_\varepsilon \gg d$$
(where l_i and l_ε are the mean free paths of electrons, corresponding to relaxation in momentum and energy, and d is the characteristic dimension of the contact). Electrons move along ballistic paths in the electric potential field, concentrated in the

area of the constriction. If the kinetic energy of the electrons is much larger than the applied bias, eV, then the trajectories of the electrons are a set of straight lines passing through the aperture (Figure 3, curves 1 and 1'). The great majority of

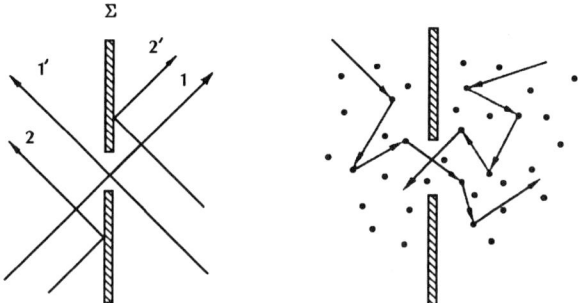

Figure 3 Trajectories of electrons in a clean contact in the form of an aperture: 1, 1' transmission; 2, 2' reflection

Figure 4 Electron trajectories in the diffusion regime for a contact in the form of an aperture

electrons are reflected from the partition (Figure 3, curves 2 and 2'). The degree of specular reflection has no influence on the characteristics of the point contact.

In impure metals for which the electron momentum mean free path is l_i, the *diffusion regime* (dirty limit) is realized [68] by satisfying the condition

$$l_i \ll d \ll \Lambda_\varepsilon \tag{1}$$

where $\Lambda_\varepsilon = \sqrt{l_i l_\varepsilon / 3}$ is the energy relaxation diffusion length of electrons (Figure 4).

The spatial dependence of the electric potential for contacts in the form of apertures has, in the clean limit, the following form:

$$V(r) = \left[1 - \frac{\Omega_o(r)}{2\pi}\right] \frac{V}{2} \, sgn\, z$$

where $\Omega_o(r)$ is the solid angle enclosing the velocities of electrons passing through the aperture and the point **r**. For a circular aperture of radius a in the dirty limit [250]

$$V(r) = \frac{V}{\pi} \arctan \xi(r) \, sgn\, z$$

Figure 5
Distribution of the potential along the z-axis for a contact in the form of an aperture:
1 - clean limit
2 - dirty limit

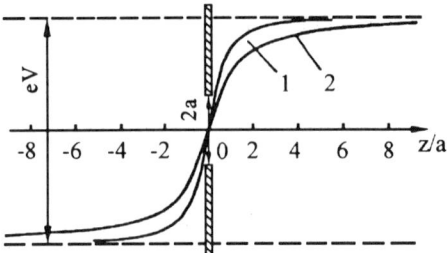

The equipotential surfaces are essentially ellipsoids of revolution with principle foci lying on diametrically opposite edges of the aperture. These surfaces coincide with surfaces of ξ = const. for the spherical coordinates ξ, η, Φ [51], where the connection to Cartesian coordinates is given by the relations

$$x = a(1 + \xi^2)^{1/2}(1 + \eta^2)^{1/2}\cos\Theta,$$
$$y = a(1 + \xi^2)^{1/2}(1 - \eta^2)^{1/2}\sin\Theta,$$
$$z = a\xi\eta.$$

In both of these regimes, drops in the potential are concentrated in regions with characteristic dimensions on the order of d near the constriction. For circular aperture models the dependence of the electrical potential on the coordinate z, measured from the center of the contact along its axis, is given by the equations [62,110]

$$V(z) = \text{sgn}\, z \, \frac{V}{2}\left(1 + \frac{a^2}{z^2}\right)^{-\frac{1}{2}} \qquad \text{(ballistic regime)}$$

$$V(z) = \text{sgn}\, z \, \frac{V}{\pi} \arctan \frac{z}{a} \qquad \text{(diffusion regime)}$$

(where V is the voltage applied to the contact and a is the radius of the point contact). The corresponding potential distribution is shown in Figure 5. The asymptotic behavior of the potential for z >> a has the form:

$$V(z) = \text{sgn}\, z \, \frac{V}{2}\left(1 - \frac{1}{2}\frac{a^2}{z^2}\right) \qquad \text{(ballistic regime)}$$

$$V(z) = \text{sgn}\, z \, \frac{V}{2}\left(1 - \frac{2a}{\pi z}\right) \qquad \text{(diffusion regime)}$$

In clean contacts the electric potential drop is concentrated within smaller intervals than in dirty contacts.

1.3 Resistance at Zero Bias

In the ballistic regime the resistance of a point contact with an aperture of the type shown (see Figure 1a) is independent of its shape and is determined by the equation [62]

$$R_0 = \frac{(2\pi \hbar)^3}{e^2 S S_F} \frac{1}{\langle \cos(\widehat{v, v_z}) \rangle_{v_z > 0}} \qquad (2)$$

where S is the area of the aperture for an arbitrary shape, S_F is the area of the Fermi surface, and $\langle \cdots \rangle_{v_z > 0}$ is the average over the half of the Fermi surface corresponding to $v_z > 0$.

For a spherical Fermi surface

$$\langle \cos(\widehat{v, v_z}) \rangle_{v_z > 0} = 1/2, \qquad S_F = 4\pi p_F^2 = \frac{3}{2} \frac{n}{p_F} (2\pi \hbar)^3$$

(n is the concentration of conduction electrons), from which

$$R_0 = \frac{4}{3} \frac{p_F}{e^2 n} \frac{1}{S}$$

or, in the circular aperture model,

$$R_0 = \frac{16}{3\pi} \frac{\rho l}{d^2}.$$

Here $\rho l = p_F/e^2 n = 3/2 N(\varepsilon_F) v_F e^2$ is the product of the resistivity and the electron mean free path, which is a characteristic constant for a given metal, and $N(\varepsilon_F)$ is the electron density of states on the Fermi surface for one spin direction.

In the diffusion regime for a contact in the shape of a unipolar hyperboloid of revolution the resistance is [11]

$$R_0 = \cot\frac{\psi}{2} \, / \, \sigma_0 d$$

or [67]

$$R_0^{-1} = 2\sigma_0 \left[b - \sqrt{b^2 - a^2} \right]$$

Here, $d = 2a$ is the size of the neck of the constriction, 2Ψ is the angle of the opening, b is the distance between the foci of the hyperboloid, and $\sigma_0 = ne^2 l_i / p_F$ is the conductivity, which is dependent on scattering by impurities. In the limiting case, $2\Psi \to \pi$, $b \to 2a$, we obtain the Maxwell formula [250]

$$R_0 = 1/\sigma_0 d$$

for the dirty aperture model. In the dirty limit the resistance depends on the shape of the aperture.

Let a contact be represented by an aperture in the form of an ellipse with semimajor and semiminor axes $\alpha = a\zeta$ and $\beta = a/\zeta$ so that the area of the ellipse is $\pi\alpha\beta$, the same as the area of a circle with radius a [110]. The coefficient $\zeta = \sqrt{\alpha/\beta}$ is the eccentricity of the ellipse. Then

$$R_0(\alpha, \beta) = R_0(a,a) C(\zeta)$$

where $R_0(a, a) = \rho/2a$, and $C(\zeta)$ is called the *form coefficient*. A graph of $C(\zeta)$ is shown in Figure 6. For $\zeta \gg 1$ it has the asymptotic value

$$C(\zeta) \approx \frac{4}{\pi\zeta} \ln(2\zeta).$$

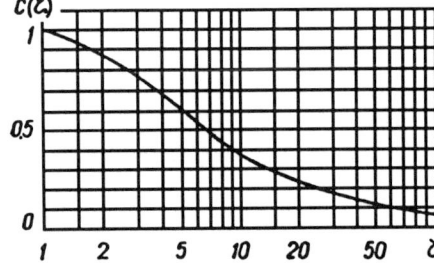
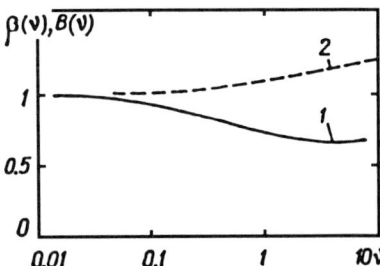

Figure 6 Form-coefficient for an elliptical dirty contact

Figure 7 Plots of the functions $\beta(v)$ for a contact in the form of an aperture [328] (1) and $B(v)$ for a contact in the form of a channel [68] (2); $\lim_{v \to \infty} \beta(v) = 9\pi^2/128 \approx 0.694$, $\lim_{v \to \infty} B(v) = 1.333$, $\beta_{min}(v) = 0.6828$ for $v = 7.6$.

In the *intermediate regime*, with an arbitrary ratio of l_i to d, the following interpolation formulas hold:
for a contact in the form of a circular aperture [328]

$$R_0 = \frac{16}{3\pi}\frac{\rho l_i}{d^2} + \beta(v)\frac{\rho}{d}, \qquad (3)$$

for a contact in the form of a long channel (with a circular cross section and specular reflective walls) [68]

$$R_0 = \frac{16}{3\pi}\frac{\rho l_i}{d^2} + B(v)\rho\frac{L}{S}.$$

In these equations $\rho = p_F/e^2 l_i$ is the resistivity due to scattering by impurities, d is the diameter of the aperture (or the channel), L is the length of the channel, and $S = \pi d^2/4$. The functions $\beta(v)$ and $B(v)$ depend on the ratio of l_i to the characteristic length l_B (Fig. 7):

$$v = l_i/l_B, \quad l_B = \begin{cases} 3\pi a/8 & \text{for an aperture,} \\ L & \text{for a channel.} \end{cases}$$

For a channel $B(v)$ has the following form [68]:

$$B(v) = \frac{4}{3}\frac{1 + \dfrac{1}{2v}\ln(1 + 2v)}{1 + \dfrac{1}{v}\left[\dfrac{1}{2v}\ln(1 + 2v) - 1\right]}.$$

Obviously, corrections based on the functions $\beta(v)$ and $B(v)$ are important only in the intermediate regime, that is, $l_i \sim d$.

In the case of a heterogeneous distribution of scattering centers, producing a semi-transparent barrier in the plane of the aperture (see Fig. 2a.), the resistance of the contact is

$$R_T = R_0/T,$$

where T is the transmission coefficient ($T \leq 1$). For modeling semitransparent barriers, a one dimensional repulsive potential is used $v(z) = H\delta(z)$ (where $\delta(z)$ is a delta function), presenting a barrier of strength $Z = H/\hbar v_F$ for an electron, and having a transmission coefficient $T = 1/(1 + Z^2)$.

If a contact is formed between differing metals (a heterocontact), with Fermi velocities v_{F1} and v_{F2}, there is a corresponding reflection of electrons from the plane of the aperture even in the absence of scatterers. This effect can be approximated by introducing an effective barrier strength [159]

$$Z_{eff} = [Z^2 + (1 - \varkappa)^2/4\varkappa]^{1/2},$$

where $\varkappa = v_{F1}/v_{F2}$.

Real metal contacts sometimes consist of several microconstrictions with parallel connections. The junction is then simply modeled as N identical, independent point contacts with resistances $R = R_0/N$.

1.4 Nonequilibrium Electron Distribution Functions

At a sufficiently high bias, eV >> kT, the momentum and energy distribution functions for electrons in the microconstriction region are in considerable nonequilibrium and are position dependent.

For the ballistic regime in the zeroth approximation (without taking inelastic collisions in the region of the contact into account) [62]

$$f(p, r) = f_0 \left\{ \varepsilon_p - \frac{eV}{2}[\eta(p, r) - (1 - \Omega_0(r)/2\pi)sgn z] \right\}. \quad (4)$$

Here $f_0 = \{\exp[(\varepsilon_p - \mu)/kT] + 1\}^{-1}$ is the equilibrium electron distribution function (the Fermi-Dirac distribution function) for energy ε_p, μ is the chemical potential, and Ω_0 is the solid angle which encloses the velocities of the electrons, **v**, corresponding to transit trajectories of type 1 or 1' in Figure 3 and passing through point **r**. The function $\eta(p,r)$ has the form

$$\eta(p, r) = \begin{cases} - sgn\, z & for \quad v \in \Omega_0(r), \\ + sgn\, z & for \quad v \notin \Omega_0(r). \end{cases}$$

For a point **r** lying in the plane of the aperture, equation (4) simplifies to:

$$f(p) = f_0 \left(\varepsilon_p + \frac{eV}{2} sgn\, v_z \right).$$

The Fermi surface, which at a temperature of T = 0 separates occupied states in **p**-space from free ones, splits in half and is moved to energies of +eV/2 for states corresponding to $v_z > 0$ and to -eV/2 for states with $v_z < 0$ (Figure 8a). For arbitrary points **r** in the vicinity of the aperture, the occupied surface also consists of two parts, with parts of the initial (undisturbed) Fermi surface being moved relative to each other by eV in energy such that the volume it envelops in **p**-space remains constant (this maintains the electrical neutrality of the metal at each point). The angular width of the protrusion (or notch) on this surface corresponds to the solid angle Ω_0, which is occupied by the aperture as seen from the given point **r** (Figure 8b).

The presence of elastic scattering centers in the region of the contact (the intermediate regime) has the result that some occupied states inside the "energized" part of the Fermi surface are found to be free, while some free energy states lying under the surface of energy $\varepsilon_F + eV/2$ for $v_z < 0$ are occupied (Figure 9).

In the limiting case of dirty contacts (the diffusion regime) the distribution of electrons in momentum space becomes nearly isotropic [67]. Nevertheless, the difference in energy between surfaces separating partially filled states from the free ones on the one side and from occupied states on the other is equal to eV, as before. This difference in energy is the "energy probe" by which point contact spectroscopy is accomplished in the diffusion regime.

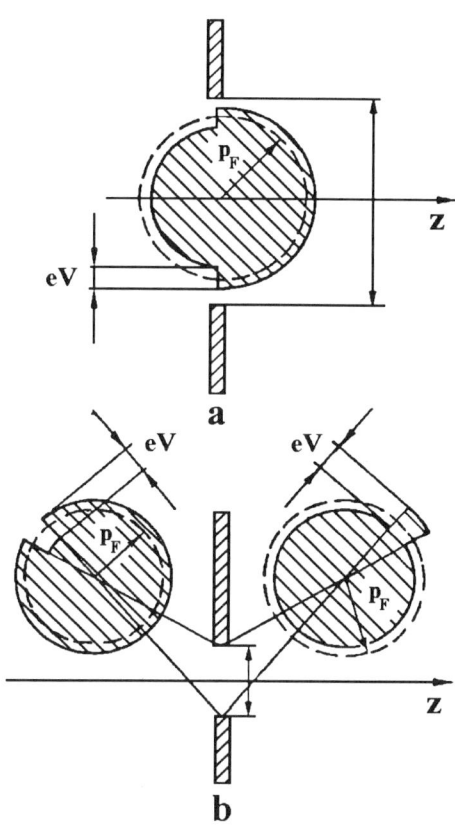

Figure 8 Momentum distribution of electrons in the clean limit:
a - in the plane of the aperture;
b - far away from it.

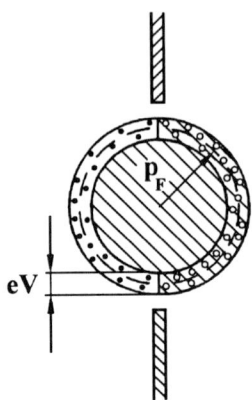

Figure 9
The electron momentum distribution in the dirty limit. All states inside the surface $\varepsilon_F - eV/2$ are completely filled while those outside $\varepsilon_F + eV/2$ are empty. States inside the energy layer of thickness eV are partially filled. The white dots symbolize free states, which would be filled in the clean limit, and the dark ones filled states which would be free. The number of filled states with $v_z > 0$ exceeds the number of states with $v_z < 0$, resulting in an electron current from left to right.

1.5 Principal Theoretical Relationships

A key theoretical relationship connects the second derivative of the point contact current-voltage characteristics $d^2I/dV^2(eV)$ with the *electron-phonon interaction point contact function*, $G(\hbar\omega)$. This connection has its simplest form in the limit $T \to 0$, which is reached when kT is much smaller than the typical energy widths of the features (extrema, breaks, etc.) in the function $G(\hbar\omega)$ [62]:

$$\frac{d^2I}{dV^2}(eV) = -\frac{\pi e^3}{\hbar \langle K \rangle} \Omega_{eff} N(\varepsilon_F) G(\hbar\omega)\Big|_{\hbar\omega = eV}. \tag{5}$$

Here, $N(\varepsilon_F)$ is the electron density of states on the Fermi surface for one direction of spin (in free electron models $N(\varepsilon_F) = mp_F/2\pi^2\hbar^3$), while the function $G(\hbar\omega)$ depends on the orientation of the axis of the contact, z, relative to the crystallographic direction in the metal. The quantity Ω_{eff} is called the *effective volume of phonon generation*, and $\langle K \rangle$ is the point contact form factor, averaged over the Fermi surface (Table 1).

At finite temperatures, *temperature broadening of the point contact spectra* occurs:

$$\frac{d^2I}{dV^2}(eV) = -\frac{\pi e^3}{\hbar \langle K \rangle} \Omega_{eff} N(\varepsilon_F) \int_{-\infty}^{\infty} G(\hbar\omega) \chi\left(\frac{\hbar\omega - eV}{kT}\right) d(\hbar\omega), \tag{6}$$

Table 1
The principal point contact characteristics for different theoretical models.

Model	Parameter			
	R_0	Ω_{eff}	$\dfrac{K(p, p')}{\langle K \rangle}$	$\langle K \rangle$
Clean aperture ($l_i \gg d$)	$\dfrac{16}{3\pi}\dfrac{\rho\, l_i}{d^2}$	$\dfrac{d^3}{3}$	$\dfrac{4 n n_z'\, \Theta(-n_z n_z')}{|n n_z' - n' n_z|}$	$\dfrac{1}{4}$
Clean channel ($l_i \gg L \gg d$)	$\dfrac{16}{3\pi}\dfrac{\rho\, l_i}{d^2}$	$\dfrac{\pi d^2}{4} L$	$2\Theta(-n_z n_z')$	$\dfrac{3\pi}{16}\dfrac{L}{d}$
Dirty aperture ($l_i \ll d$)	$\dfrac{\rho}{d}$	$4\dfrac{d^3}{3}\left(\dfrac{l_i}{d}\right)^2$	$\dfrac{3}{8}\left[(n_z - n_z')^2 + (\mathbf{n} - \mathbf{n}')^2\right]$	$\dfrac{3\pi}{16}\dfrac{l_i}{d}$
Dirty channel ($l_i \ll L \gg d$)	$\rho\dfrac{4L}{\pi d^2}$	$\dfrac{\pi d^2}{3} L \left(\dfrac{l_i}{d}\right)^2$	$\dfrac{3}{2}(n_z - n_z')^2$	$\dfrac{3\pi}{16}\dfrac{l_i}{d}$

where the *temperature broadening function* $\chi(T)$ corresponds identically to the function in the theory of inelastic tunneling spectroscopy [73] and has the form

$$\chi(y) = \frac{1}{kT}\frac{d^2}{dy^2}\left(\frac{y}{e^y - 1}\right), \qquad (7)$$

or

$$\chi(y) = \frac{1}{kT} e^y \frac{(y - 2)e^y + y + 2}{(e^y - 1)^3}. \qquad (8)$$

Extremely narrow peaks in the point contact spectra become bell-shaped curves at finite temperatures and have a width at half maximum of 5.44 kT. If the integral intensity of a sharp δ-shaped peak is adopted as a unit, then the height of a temperature broadened peak is 1/6 kT.

In practice, the following form of equation (5) is often used, for which it is assumed that the contact is analogous to a clean circular aperture in form (see Figure 1a), and the Fermi surface is spherical:

$$\frac{d(\ln R)}{d}V \approx R_0 \frac{d^2 I}{dV^2}(eV) = -\frac{32}{3}\frac{ed}{\hbar v_F}G(\hbar\omega)\bigg|_{\hbar\omega = eV}. \qquad (9)$$

1.6 The Electron-Phonon Interaction Point Contact Functions

The EPI point contact function, $G(\omega)$, which enters into expressions (5), (6), and (9), has the following form [62]:

$$G(\omega) = \frac{(2\pi\hbar)^{-3}}{\oint dS_p/v} \oint_{FS} \frac{dS_p\, dS_{p'}}{v\, v'} \sum_s |M_{p-p',s}|^2 K(\mathbf{p},\mathbf{p}')\, \delta(\omega - \omega_{p-p',s}). \qquad (10)$$

The integral here is taken in the quasi-momentum of the electrons $\mathbf{p}$ and $\mathbf{p}'$, lying on the Fermi surface and having corresponding velocities $v(\mathbf{p})$ and $v(\mathbf{p}')$. For an electron in state $\mathbf{p}$ the probability of being scattered into state $\mathbf{p}'$ by a phonon with energy $\hbar\omega_{p-p'}$ and branch number s equals $\wp_{p-p',s} = \frac{2\pi}{\hbar}|M_{p-p',s}|^2$, where $|M_{\mathbf{p}-\mathbf{p}',s}|$ is the modulus of the transition matrix element for $\mathbf{p} \to \mathbf{p}'$. The summation in s implies summing over all branches of the phonon spectrum. The $K(\mathbf{p},\mathbf{p}')$ factor in equation (10) takes into account the influence of the geometric form of the point contact and its cleanliness on electron-phonon scattering processes in the region of the constriction. The K-factors for various contact models are shown in Table 1. The K-factor determines the sensitivity of point contact functions to the direction of the contact axis relative to the crystallographic direction in the metal (anisotropy of the point contact spectra). In general, anisotropy of the point contact spectra depends on the anisotropy of the Fermi surface, as well as that of the phonon spectrum.

In the particular case of a spherical Fermi surface (the free electron model) the anisotropy of the point contact spectra is completely determined by the anisotropy of the phonon spectrum. For the clean circular aperture model the K-factor can be written in the form [130]

$$K(\Theta,\alpha) = \frac{2}{\pi \sin\Theta} \int_{\varphi_0}^{\pi/2} \frac{s^2 z^2 - c^2(1-z)^2 \cos^2\varphi}{(\cos^2\varphi + z^2\sin^2\varphi)^{1/2}}\, d\varphi,$$

where $\varphi_0 = \cos^{-1}\left(sz/c\sqrt{1-z^2}\right)\Theta\left(c\sqrt{1-z^2} - sz\right)$, $\Theta(\chi)$ is the theta function, $z = \cos\alpha$, $s = \sin(\Theta/2)$, $c = \cos(\Theta/2)$, and $\sin(\Theta/2) = q/2p_F$. The scattering angle, Θ, is given by the modulus of the transferred wave vector $|\mathbf{q}| = |\mathbf{p} - \mathbf{p}'|$ (see "a" in the inset of Figure 10) and determines the effective scattering process. The angular selectivity of the method of point contact spectroscopy is determined by the dependence of the K-factor on the angle α, between the vector $\mathbf{q}$ and the axis of the contact (see "b" in the inset of Figure 10). The dependence $K(\alpha)$ for different angles Θ, as shown in Figure 10, is characteristic of the directionality in the point contact spectroscopy method for studies of anisotropic metals.

If a clean aperture has the form of an ellipse with a semi-axis ratio ζ, then for $\zeta \ll 1$ the K-factor is [85]

$$K_{ell}(v,v') = \frac{\zeta^{1/2}}{\sqrt{W_x^2 + \zeta^2 W_y}} |v_z| \Theta(-v_z v_z'),$$

where $W = (v_z v' - v_z' v)/v_z'$ and $\Theta(x)$ is the theta function. In this case the effective volume of phonon generation is

$$\Omega_{eff}^{ell} = \frac{8}{3}\left(\frac{S}{\pi}\right)^{3/2}.$$

The form factor of a rectangular aperture with sides $a\sqrt{\vartheta}$ and $a/\sqrt{\vartheta}$ (ϑ is the ratio of the sides of the rectangle) and area $S = a^2$ can, in approximately isotropic metals with spherical Fermi surfaces, be expressed in the form [245]

$$K_{rec}(v,v') = K_0(v,v') \Omega_{eff}^{rec},$$

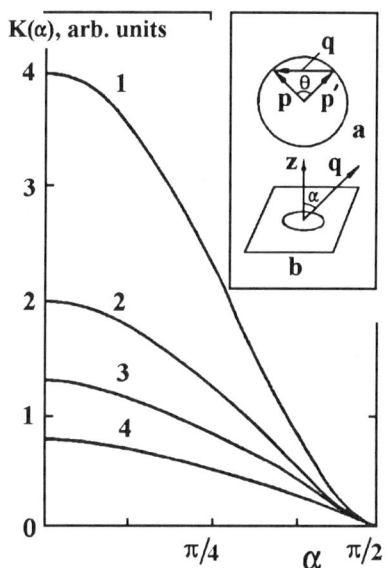

Figure 10 $K(\alpha)$ for various values of Θ:
1 - $\Theta = 0.99\pi$, 2 - 0.98π, 3 - 0.97π, 4 - 0.95π
(In the inset: $\mathbf{q}$ is the scattering vector between states $\mathbf{p}$ and $\mathbf{p}'$ on the Fermi surface, as shown in (a). Its orientation with respect to the contact axis is shown in (b).)

where $K_0(v, v')$ is the normalized form factor for circular apertures:

$$K_0(v,v') = \frac{4\Theta(-v_z v_z')}{|v_z'v - v_z v'|}, \quad \Omega_{eff}^{rec} = S^{3/2} C_{rec}(\vartheta),$$

$$C_{rec}(\vartheta) = \frac{1}{4\pi}[\Phi(\vartheta) + \Phi(\vartheta^{-1})],$$

and

$$\Phi(\vartheta) = \sqrt{\vartheta} \ln\left(\sqrt{1 + \vartheta^{-2}} + \vartheta^{-1}\right) - \frac{\vartheta^{3/2}}{3}\left(\sqrt{1 + \vartheta^{-2}} - 1\right).$$

For a square ($\vartheta = 1$), $C_{rec}(\vartheta=1) = 0.1183$, whereas for a circular aperture $C_{cir} = 0.1197$ and $\Omega_{eff} = S^{3/2}C_0$, where $S = \pi a^2$. From the dependence of $C_{rec}(\vartheta)$ (Figure 11) it follows that for contacts in the form of rectangular apertures with a ten-fold ratio of the sides the intensity of the electron-phonon interaction function calculated in the circular aperture model is underestimated by approximately 30%.

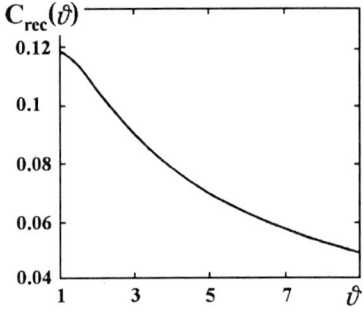
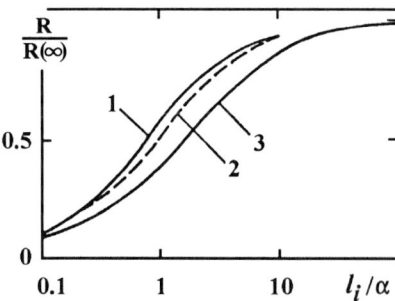

Figure 11 The form coefficient for a contact in the shape of a rectangular aperture.

Figure 12 The dependence of the resistance of a point contact in the shape of a circular aperture on the parameter l_i/α:
1 - data of reference [328]; 2 - [246]; 3 - [67].

For a point contact in the form of a one-sheet hyperboloid of revolution (see Figure 1b) the dependence of the K-factor on the cleanliness parameter $v = l_i/d$ (where l_i is the average elastic scattering length and d is the diameter of the neck of the hyperboloid) can only be calculated numerically using interpolation procedures, as shown in reference [67]. In the special case of a circular aperture in a thin partition this procedure leads to the dependence of the zero bias resistance on v shown in Figure 12 (curve 2). Curve 1 in Figure 12 is the result obtained by interpolation of equation (3).

The K-factor of a long channel (L >> d) of arbitrary purity, calculated using interpolation procedures, has the form [67]

$$K(\mathbf{n},\mathbf{n}') = \frac{9\pi}{32} \frac{L}{d} \frac{v}{\gamma(v)} \left\{ (n_z - n_z')^2 + n_z^2 \exp\frac{1}{v|n_z|} + \right.$$

$$n_z'^2 \exp\left(-\frac{1}{v|n_z'|}\right) + 4v(n_z' - n_z) \left[n_z^2 \exp\left(-\frac{1}{2v|n_z|}\right) \sinh\frac{1}{2vn_z} - \right.$$

$$n_z'^2 \exp\left(-\frac{1}{2v|n_z|}\right) \sinh\frac{1}{2vn_z'} + 4v \frac{(n_z n_z')^2}{(n_z - n_z')} \times$$

$$\left. \left. \exp\left(-\frac{1}{v}\frac{|n_z| + |n_z'|}{2|n_z n_z'|}\right) \sinh\frac{n_z' - n_z}{2v n_z n_z'} \right] \right\},$$

where

$$\gamma(v) = 3 \int_0^1 t^2 \left[1 - \exp\left(-\frac{1}{2vt}\right)\right] dt, \quad \mathbf{n} = \mathbf{v}/v_F, \quad \mathbf{n}' = \mathbf{v}'/v_F .$$

If the scattering of electrons is considered to occur essentially only at 180° (this is called the **(p,− p)** model, proposed in reference [68]), then for models of a channel with arbitrary purity, circular cross section, and adjoining clean banks, the result is [67]

$$K_{chan} = \frac{4v^2|n_z n_z'|[1 - v^{-1} + (2v^2)^{-1}\ln(1 + 2v)]^{-1}}{(1 + 2v|n_z|)(1 + 2v|n_z'|)} \left\{ K_0(\mathbf{n},\mathbf{n}') + \frac{3\pi}{32}\frac{L}{d}\frac{[n_z' - n_z + 4vn_z'|n_z|\Theta(-n_z n_z')]^2 - \frac{1}{3}(|n_z| - |n_z'|)^2}{|n_z n_z'|(1 + 2v|n_z|)(1 + 2v|n_z'|)} \right\}$$

Here $K_0(\mathbf{n}, \mathbf{n}') = |n_z n_z'|\Theta(-n_z n_z')/|n_z'\mathbf{n} - n_z\mathbf{n}'|$ is the form factor of a clean circular aperture and $v = l_i/L$. The mean effective value of the K-factor in such a model is

$$\langle K_{chan} \rangle = \frac{\pi}{32}\frac{L}{d}\frac{24v^3 + 20v^2 + (1 + v)\ln(1 + 2v)[\ln(1 + 2v) - 12v]}{(1 + 2v)[2v^2 - 2v + \ln(1 + 2v)]} + \frac{4}{\pi}v^4[2v^2 - 2v + \ln(1 + 2v)]^{-1} T(v) , \quad (11)$$

where

$$T(v) = \int_0^1 dx \int_0^1 dy \int_0^\pi d\varphi \frac{(xy)^2}{(1 + 2vx)(1 + 2vy)}(x^2 + y^2 - 2x^2y^2 - 2xy\sqrt{1 - x^2}\sqrt{1 - y^2}\cos\varphi)^{1/2}.$$

Table 2 shows the results of a numerical calculation of equation (11) for the dependence of the average form factor for a channel on the parameter v, with $L = d$. The contributions by the channel and banks are listed separately, along with the quantity $\langle K \rangle$, normalized to its value at $v = \infty$ (Figure 13).

A simplified formula for $\langle K_{chan} \rangle$ from the model of 180° scattering (the contribution of the banks is ignored), assuming a long circular channel of arbitrary

Table 2
The form factor of a channel, averaged over the Fermi surface, for different ratios of the momentum mean free path to the channel length.

$2l_i/L$	$\langle K \rangle$	Channel	Bank	$\dfrac{\langle K \rangle}{\langle K(\infty) \rangle}$
∞	0.8391	0.5891	0.25	
100	0.8047	0.5654	0.2393	0.9590
90	0.8025	0.5634	0.2391	0.9564
80	0.7998	0.5611	0.2387	0.9532
70	0.7964	0.5581	0.2383	0.9491
60	0.7922	0.5544	0.2378	0.9441
50	0.7865	0.5410	0.2370	0.9373
40	0.7885	0.5426	0.2359	0.9278
30	0.7662	0.5322	0.2340	0.9131
20	0.7447	0.5143	0.2304	0.8875
10	0.6938	0.4731	0.2207	0.8268
9	0.6843	0.4655	0.2188	0.8155
8	0.6730	0.4567	0.2163	0.8020
7	0.6594	0.4461	0.2133	0.7858
6	0.6426	0.4332	0.2094	0.7658
5	0.6213	0.4169	0.2044	0.7404
4	0.5929	0.3957	0.1972	0.7066
3	0.5528	0.3662	0.1866	0.6588
2	0.4903	0.3214	0.1689	0.5849
1	0.3785	0.2404	0.1381	0.4511
0.9	0.3546	0.2282	0.1264	0.4226
0.8	0.3344	0.2146	0.1198	0.3985
0.7	0.3117	0.1996	0.1121	0.3715
0.6	0.2861	0.1827	0.1034	0.3410
0.5	0.2528	0.1635	0.0893	0.3013
0.4	0.2228	0.1414	0.0814	0.2655
0.3	0.1828	0.1562	0.0672	0.2179
0.2	0.1348	0.0849	0.0499	0.1606
0.1	0.0756	0.0475	0.0281	0.0901
0.09	0.0689	0.0432	0.0257	0.0821
0.08	0.0620	0.0389	0.0231	0.0739
0.07	0.0550	0.0345	0.0205	0.0655
0.06	0.0477	0.0299	0.0178	0.0569
0.05	0.0403	0.0252	0.0151	0.0480
0.04	0.0327	0.0246	0.0122	0.0389
0.03	0.0249	0.0156	0.0093	0.0296
0.02	0.0168	0.0105	0.0063	0.0200

purity, is more convenient for analytical calculations. It has the form

$$\langle K_{chan} \rangle = \frac{1 - v\ln(1+2v) + (1+2v)^{-1}}{1 - v^{-1} + (2v)^{-1}\ln(1+2v)}. \quad (12)$$

Here, $\langle K_{chan} \rangle$ is normalized to its value at $v = \infty$. Comparison of the normalized dependences (11) and (12) shows that, to the resolution attainable in Figure 13, they are indistinguishable. Since the dependence of the magnitude of point contact spectra on v is proportional to $\langle K \rangle$, equations (11) and (12) reflect a decrease in magnitude with increasing contamination in the region of the constriction.

From Table 1 it follows that for circular apertures models the magnitude of the point contact spectra (that is, the absolute value of d^2I/dV^2) is proportional to $R_0^{-3/2}$, while for other models it is replaced by the inverse proportional dependence: $d^2I/dV^2(V) \propto R_0^{-1}$ (under the condition that ρ and l_i are constants). This property is used in practice to verify the conformance of real contacts to models of (circular) clean apertures [168, 170].

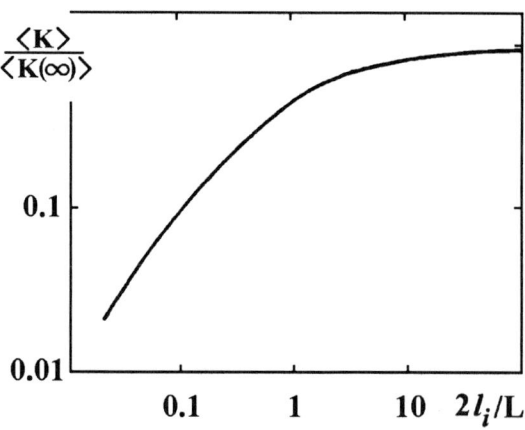

Figure 13
Dependence of the normalized form factor for a point contact in the form of a channel on the parameter $2l_i/L$ [67].

1.7 Heterocontacts

A *heterocontact* is a point contact formed between dissimilar electrodes. The differences can pertain to the electron and phonon spectra, the degree of purity (differences in the elastic mean free path, l_i) and the geometric forms of the contacting electrodes. Some possible models for heterocontacts are shown in Figure 14.

For phonon spectroscopy in heterocontacts the following characteristics are observed [126].

1. The point contact spectrum of a heterocontact is always symmetric relative to the polarity of the applied voltage.

2. Differences in the elastic scattering length l_i in the electrodes do not have an influence on the intensity ratios of their point contact spectra. Contamination in one of the metals lowers the magnitude of its contribution to the point contact spectrum as well as that of the other, clean, metal.

3. Geometrical asymmetry of a contact results in a difference in the effective volumes of phonon generation in each of the metals, which then causes a difference in the partial intensities of their point contact spectra. This is equally true for clean and dirty metals.

4. In a metal with a large value of p_F the relative phase space of states filled due to the nonequilibrium is smaller due to reflection of some of the electron trajectories from the interface. This leads to a relative increase in the inelastic relaxation time, τ_ε. Since the intensity of the point contact spectrum is proportional to $d/v_F\tau_\varepsilon$, the partial contribution by a good metal (that is a metal with large values of p_F and v_F) to the point contact spectrum proves to be smaller.

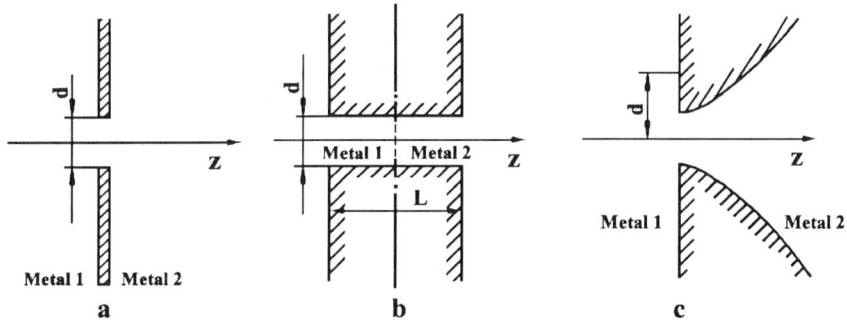

Figure 14 Models of heterocontacts
a,b - clean contacts in the form of an aperture and a channel: c - a contact of clean (1) and dirty (2) metals (In the latter case the characteristic dimension of the contact, d, is determined using equation (19).).

5. If $p_{F1} \gg p_{F2}$, the point contact spectrum of the metal with the larger value of p_F is drawn into narrow peaks, corresponding to frequencies of phonons having momenta differing from $2p_{F1}e_z$ (where e_z is the unit vector oriented along the axis of the contact) by an inverse lattice vector $\mathbf{Q}$. The magnitude and width of a peak depend on the component of the phonon velocity perpendicular to the contact axis. Impurities in the metal broaden these peaks and in the dirty limit they transform to the standard point contact type, being nonzero for the entire range of phonon

frequencies. A general relationship in point contact spectroscopy theory, valid when $eV \ll \varepsilon_F^j$ for all heterocontacts, has the form [126]

$$\frac{1}{R}\frac{dR}{dV} = \frac{32}{3}\frac{ed}{\hbar} \sum_{j=1}^{2} \frac{1}{v_F^j} \int_0^\infty \frac{d\omega}{kT} S\left(\frac{\hbar\omega - eV}{kT}\right) G_j(\omega), \quad (13)$$

where

$$G_j(\omega) = \frac{(2\pi\hbar)^{-3}}{\oint dS_p/v} \oint_{FS} \frac{dS_p dS_{p'}}{vv'} \sum_{Q_s} |M_{p-p'+Q,s}|^2 K_j(p,p') \times$$

$$\delta(\omega - \omega^j_{p-p'+Q,s}), \quad (14)$$

$G_j(\omega)$ is the EPI point contact function of metal j, taking into account the scattering of electrons at the interface. The summation in equation (14) is carried out over all branches of the phonon spectrum, s, and inverse lattice vectors, **Q**. The function $S(\chi)/kT$ in expression (13) is the standard temperature broadening function (see equations (7) and (8)) and d is the characteristic dimension of the contact (see Figure 14). The form factor $K_j(\mathbf{p}, \mathbf{p}')$ differs from the K-factor of the monocontact (that is, a contact between identical electrodes) and for the most important specific cases is given by the following expressions.

1. *A clean heterocontact* with a δ-function potential barrier $U(z) = U_0\delta(z)$ at the interface:

$$K_j(p,p') = \frac{D(p_j = p)D(p_j = p')}{4\langle|n_z|D\rangle_j} K_{0,j}(p,p').$$

Here, $K_{0,j}(\mathbf{p}, \mathbf{p}')$ is the K-factor of a clean monocontact, and

$$\langle \cdots \rangle_j = \frac{\oint_{FS_j} \frac{dS_p}{v_j}(\cdots)}{\oint_{FS_j} \frac{dS_p}{v_j}}.$$

The probability of electron transmission through the interface equals

$$D(p, p') = \frac{4\hbar^2 |v_z \bar{v}_z| \Theta(p_{F2} - |p_{\|,1}|) \delta_{p_\| \bar{p}_\|}}{\hbar^2 (|v_z| + |\bar{v}_z|)^2 + 4U_0},$$

where **p**, **v** and $\bar{\mathbf{p}}$, $\bar{\mathbf{v}}$ are the momenta and velocities of the incident and exiting electrons, $\mathbf{p}_\|$ is the component of the momentum parallel to the axis of the contact, $\delta_{\alpha,\beta}$ is the Kronecker symbol, and $\Theta(\chi)$ is the theta function.

2. *A contact of two dirty metals* ($l_i^j \ll d$):
a) when scattering on the boundary is absent (a contact of identical metals with differences in purity), $D = 1$, $p_{F1} = p_{F2}$

$$K_j(p, p') = \frac{l_i^j l_i^{\bar{j}}}{d(l_i^j + l_i^{\bar{j}})} K_{0,j}(p, p') ; \qquad (15)$$

b) in the case of strong scattering on the boundary ($D \ll 1$):

$$K_j(p, p') = \frac{2}{3} \frac{l_i^j l_D^{\bar{j}}}{l_i^j l_i^{\bar{j}} + l_i^j l_D^{\bar{j}} + l_i^{\bar{j}} l_D^j} K_{0,j}(p, p') ; \qquad (16)$$

c) in the case of very strong scattering on the boundary ($l_D \ll l_i$):

$$K_j(p, p') = \frac{2}{3} \frac{l_D^j}{l_i^j} K_{0,j}(p, p').$$

In formulas (15) and (16) the following notation has been introduced: $j, \bar{j} = 1, 2; \; j \neq \bar{j}$;

$$l_D^j = \frac{\pi d}{2} \frac{\left[\langle D \rangle_{\bar{j}} (1 - \langle |\bar{n}_z| D \rangle_{\bar{j}}) + \langle D \rangle_{\bar{j}} \langle |n_z| D \rangle_j\right] \langle |\bar{n}_z| D \rangle_j}{(1 - \langle |n_z| D \rangle_{\bar{j}})(1 - \langle |\bar{n}_z| D \rangle_{\bar{j}}) - \langle |n_z| D \rangle_{\bar{j}} \langle |n_z| D \rangle_j}.$$

In the limiting cases, for which formulas (15) and (16) apply, the expression for l_D^j simplifies to:

$$l_D^j = \begin{cases} \infty & \text{for } D = 1, \\ \dfrac{\pi d}{2} \langle D \rangle_j & \text{for } D \ll 1. \end{cases}$$

Since for a dirty heterocontact $\langle D \rangle_1 / \langle D \rangle_2 \approx (p_{F2} / p_{F1})^2$, the ratio of partial intensities in the point contact spectra is

$$\frac{[d(\ln R)/dV]_1}{[d(\ln R)/dV]_2} \approx \frac{V_{F2}}{V_{F1}} \left(\frac{p_{F2}}{p_{F1}} \right)^2.$$

3. *A contact of clean (j = 1) and dirty (j = 2) metals:*

$$K_1(p, p') = K_{0,1}(p, p') \frac{l_1^{(2)}}{2\pi d} \begin{cases} 0.43(1 + 2|n_z|)(1 + 2|n_z'|), & D = 1, \\ \dfrac{D(p_1 = p) D(p_1 = p')}{\langle D \rangle_2 \langle |n_z| \rangle_1}, & D \ll 1, \end{cases} \quad (17)$$

$$K_2(p,p') = K_{0,2}(p,p') \begin{cases} 1.14 & \text{for } D = 1, \\ 0.67 \dfrac{\langle D \rangle_2}{\langle |n_z| D \rangle_2} & \text{for } D \ll 1. \end{cases} \quad (18)$$

In equations (15) and (16) $K_{0,j}(p,p')$ is the form factor for monocontacts of the corresponding dirty metals. In equations (17) and (18), $K_{0,1}(p, p')$ and $K_{0,2}(p, p')$ are the form factors for monocontacts of clean metal 1 and dirty metal 2. The case of $D = 1$ corresponds to identical metals ($p_{F1} = p_{F2}$) having, however, different impurity levels.

An electrode in the form of a hyperboloid of revolution (Figure 14c) can be dealt with only in the dirty limit. In this case its characteristic dimension [67] is:

$$d = \frac{\sigma_0 R_0 S}{\pi} + \frac{1}{4\sigma_0 R_0}, \quad (19)$$

where R_0 is the resistance of a symmetric point contact in the form of a hyperboloid of revolution, S is the cross sectional area of the neck, and σ_0 is the electrical conductivity of the metal.

1.8 Two-phonon Processes

Multi-phonon processes can be understood as processes of consecutive emission (or absorption) of n phonons by an electron passing through the region of the constriction. The magnitude of peaks in the point contact spectra due to these processes is proportional to $(d/l_\varepsilon)^n$.

Additions to the second derivative of the current-voltage characteristics of a point contact due to two-phonon processes have the form (for the model of a circular channel of length L and diameter d (L >> d)) [61]

$$\left(\frac{d^2 I}{dV^2}\right)^{(2)} = -\frac{32e}{3R_0 \omega_0^2} \int_0^{eV} f_{pc}(\omega, eV - \omega) d\omega,$$

where R_0 is the resistance for V = 0, $\omega_0 = v_F/d$,

$$f_{pc}(\omega_1, \omega_2) = \frac{(2\pi\hbar)^{-3}}{\oint_{FS} dS_p/v} \oint_{FS} \frac{dS_p}{v} \frac{dS_{p'}}{v'} \frac{dS_{p''}}{v''} K_f(p, p', p'') \times$$

$$|M_{p,p'}|^2 |M_{p',p''}|^2 \delta(\omega_1 - \omega_{p'-p}) \delta(\omega_2 - \omega_{p''-p'}),$$

and $\quad K_f(p, p', p'') = \frac{3}{8} \frac{L}{d^2} \frac{p_z}{|p'_z|} \Theta(-p_z p'_z) \Theta(p'_z p''_z).$

If it is assumed that the square of the electron-phonon interaction matrix element is independent of $q = (p' - p)$ and the phonon spectrum is Einsteinian (ω_q = const.), then $f_{pc}(\omega_1, \omega_2) = G(\omega_1)G(\omega_2)$. The calculation of a two-phonon correction to the point contact spectrum is difficult in the circular aperture model because of the divergence of the integrals at large distances from the aperture.

Two-phonon processes cause peaks in the point contact spectra for energies equal to double the magnitude of the characteristic phonon energies, as well as for other energy combinations.

1.9 Comparison of Electron-Phonon Interaction Point Contact Functions to Related Functions

If the energy of a phonon is $\hbar\omega$, and its wave vector is q, then the *phonon density of states* is expressed by the equation

$$F(\omega) = \frac{1}{N} \sum_s \int_{BZ} \frac{d^3q}{(2\pi\hbar)^3} \delta(\omega - \omega_{q,s}). \quad (20)$$

Here, N is the number of ions per unit volume, $\omega_{q,s}$ is the phonon dispersion relation corresponding to branches of the spectrum with number s, and the integral is over the first Brillouin zone.

The *EPI thermodynamic function* (the Eliashberg function)

$$g(\omega) = \frac{(2\pi\hbar)^{-3}}{\oint_{FS} dS_p/v} \oint_{FS} \frac{dS_p \, dS_{p'}}{vv'} \sum_s |M_{p-p',s}|^2 \delta(\omega - \omega_{p-p',s})$$

may conventionally be represented in the form

$$g(\omega) = \alpha^2(\omega) F(\omega),$$

where $\alpha^2(\omega)$ is the square of the modulus of the EPI matrix element averaged over the Fermi surface. In particular, for a spherical Fermi surface and a local pseudopotential with a Fourier component $W(q)$ the following equation holds [166]:

$$g(\omega) = -\frac{N(\varepsilon_F)}{8\pi p_F^2} \sum_s \int_{q<2p_F} \frac{d^3q}{q} \frac{|W(q)|^2 (q\,\varepsilon(q,s))^2}{2MN\omega} \delta(\omega - \omega_{q,s}). \quad (21)$$

Here $N(\varepsilon_F)$ is the density of states on the Fermi surface for a single orientation of spin, $\varepsilon(p,s)$ is the polarization vector of a phonon with wave vector q for branch s, M is the ionic mass of the metal, and N is the number of metal ions per unit volume. Apart from the numerical coefficient of equation (21) being different from (20), the integrand factor depends on the momentum, $q = (p - p')$, transferred in the process of electron-phonon scattering. When the integral in equation (21) is carried out, vectors q, reduced to the first Brillouin zone for the value q and outbound within its limits, are used.

The dependence of the electrical resistivity of the metal on temperature, due to the scattering of electrons by phonons, can be represented by

$$\rho_{ph}(T) = \frac{2\pi m}{ne^2} \int_0^\infty G_{tr}(\omega) \Psi\left(\frac{\hbar\omega}{kT}\right) d\omega, \qquad (22)$$

where m, n are the effective mass and concentration of conduction electrons, and

$$\Psi(\chi) = \frac{\chi}{2} \sinh^2\frac{\chi}{2}.$$

The *EPI transport function* $G_{tr}(\omega)$ depends on the orientation of the current density vector (parallel to the z axis) relative to the crystallographic directions in the sample:

$$G_{tr}(\omega) = \frac{(2\pi\hbar)^{-3}}{\oint dS_p/v} \oint_{FS} \frac{dS_p\, dS_{p'}}{vv'} \sum_s |M_{p-p',s}|^2 K_{tr}(p,p') \delta(\omega - \omega_{p-p',s}), \qquad (23)$$

where the K-factor is

$$K_{tr}(p, p') = \frac{3}{2}(p_z^2 - p_z'^2)/p_F^2.$$

Averaging over all orientations of the z axis occurs for polycrystalline samples, with the result that the resistivity becomes isotropic. It is described by using equation (22) with an EPI transport function having the form of (23), but with an isotropic K-factor

$$K(p, p') = 2(1 - \cos\Theta)$$

where Θ is the angle between the vectors **p** and **p'** (the scattering angle). For the normalized EPI transport function, $g_{tr}(\omega)$, the K-factor has the form

$$\eta_{tr}(\Theta) = K(p, p')/\langle K \rangle = (1 - \cos\Theta).$$

Here $\langle K \rangle$ is the K-factor averaged over all angles, Θ.

In order to compare the EPI point contact function with other related functions, it is convenient to normalize it by the K-factor averaged over the Fermi surface, as

shown in Table 1. In this way the *normalized EPI point contact function*,
$$g_{pc}(\omega) = G(\omega)/\langle K \rangle,$$
is formed, where

$$\langle K \rangle = \oint_{FS} \frac{dS_p \, dS_{p'}}{v \, v'} K(v, v') \bigg/ \oint_{FS} \frac{dS_p \, dS_{p'}}{v \, v'}.$$

Both $G(\omega)$ and $g_{pc}(\omega)$ are dimensionless quantities.

In the model of an isotropic (polycrystalline) metal the K-factor for a contact in the form of a clean circular aperture is expressed as:

$$K(\Theta) = \frac{1}{8}\left(1 - \frac{\Theta}{\tan\Theta}\right).$$

In such a model the normalized K-factor (Figure 15, curve 3) [321] is

$$\eta_{pc}(\Theta) = \frac{K(\Theta)}{\langle K \rangle} = \frac{1}{2}\left(1 - \frac{\Theta}{\tan\Theta}\right).$$

Substitution of the $\eta_{pc}(\Theta)$-factor into formula (10) results in the isotropic EPI point contact function

$$g_{pc}(\omega) = \frac{(2\pi\hbar)^{-3}}{\oint_{FS} dS_p/v} \oint_{FS} \frac{dS_p \, dS_{p'}}{v \, v'} \sum_s |M_{p-p',s}|^2 \eta_{pc}(\Theta) \delta(\omega - \omega_{p-p',s}).$$

In the series of normalized isotropic EPI functions,
$$g(\omega) \rightarrow g_{tr}(\omega) \rightarrow g_{pc}(\omega)$$

only the integrand factor, dependent on the scattering angle, changes

$$1 \rightarrow (1 - \cos(\Theta)) \rightarrow (1 - \Theta/\tan\Theta)/2.$$

Graphs of these functions are shown in Figure 15.

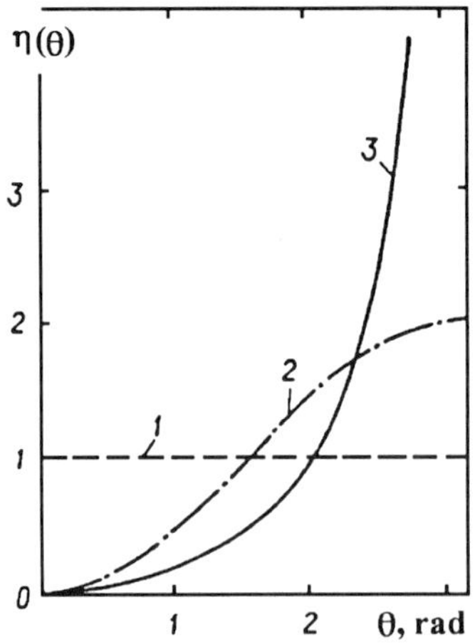

Figure 15 Dependence of the isotropic form factors of the thermodynamic (1), transport (2) and point contact (3) EPI functions on the scattering angle, Θ.

The EPI functions allow us to determine the renormalization of important characteristics of the metal in the resulting EPI:

$$m^* = (1 + \lambda)m^*_{zone}, \quad v = (1 + \lambda)v_{zone}, \quad N = (1 + \lambda)N_{zone}$$

where m^*_{zone}, v_{zone}, and N_{zone} are the effective mass, velocity and density of electron states in the absence of an EPI, m^*, v, and N are those quantities in the presence of an EPI, the normalization coefficient equals $(1 + \lambda)$, and different integral EPI parameters, λ, are related to the corresponding EPI functions

$$\lambda = 2 \int_0^\infty g(\omega)\omega^{-1}d\omega,$$

$$\lambda_{tr} = 2 \int_0^\infty g_{tr}(\omega)\omega^{-1}d\omega,$$

$$\lambda_{pc} = 2 \int_0^\infty g_{pc}(\omega)\omega^{-1}d\omega.$$

1.10 Background

The majority of point contact spectra contain a background, which is most evident for $eV > \hbar\omega_{max}$ (ω_{max} is the maximum frequency of the phonon spectrum), where the second derivative of the current-voltage characteristics is different from zero, while the EPI function identically equals zero (Figure 16). The background appears as a consequence of higher order electron-phonon scattering processes (multi-phonon processes, scattering of electrons on nonequilibrium phonons, etc.). Several empirical methods for calculating the background have been proposed. All of them should ensure the agreement (to within experimental error) of EPI point contact functions obtained from the spectra of different contacts made using the same metal. The resulting curve $g_{pc}(\omega)$ is practically independent of the method of background calculation if the relative background level (parameter $\gamma = B/A$) is less than 0.3.

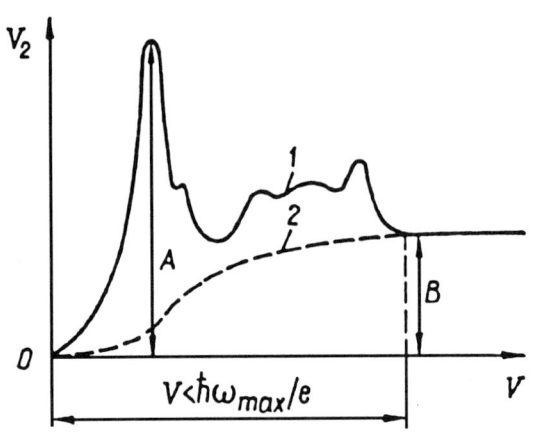

Figure 16
A point contact spectrum with a background:
1 - dependence of the voltage of the modulation signal second harmonic V_2 on the voltage on the sample, proportional to the second derivative of the current-voltage characteristics;
2 - a conjectured dependence of the background (the coefficient γ = B/A characterizes the relative background level for $eV \geq \hbar\omega_{max}$).

In the energy interval $0 < eV < \hbar\omega_{max}$ the background takes the form of one of the following functions:

$$B_1 = C_1 \, (\omega/\omega_{max}) \qquad [220],$$

$$B_2 = C_2 \, \tanh^2\left(\frac{3}{2}\frac{\omega}{\omega_{max}}\right) \qquad [218],$$

$$B_3 = C_3 \int_0^\omega g_{pc}(\omega')d\omega' \qquad [132],$$

$$B_4 = C_4 \int_0^\omega g_{pc}(\omega')\frac{d\omega'}{\omega'} \qquad [337].$$

(24)

Functions $B_3(\omega)$ and $B_4(\omega)$ are based on theoretical models, while $B_1(\omega)$ and $B_2(\omega)$

are purely empirical dependences. Constants C_{1-4} are calculated so that the shapes of the functions $B_{1-4}(\omega)$ coincide with the point contact spectrum for $eV > \hbar\omega_{max}$. Functions $B_{3-4}(\omega)$, which contain the unknown EPI point contact function, are calculated using iteration procedures. For tabulation of this atlas the background was calculated using functions of the form $B_4(\omega)$.

1.11 Point Contact Spectroscopy of Non-phonon Excitations

If non-phonon branches of the energy spectrum of quasiparticle excitations, interacting with conduction electrons, exist in a metal, then the scattering of electrons on these excitations will also be evident in the features of the point contact spectra. For metals in magnetically ordered states, characteristics dependent on electron-magnon interactions are superimposed on peaks caused by the EPI [7]. This is also the case for nearly localized electronic excitations (excitons), of importance in rare-earth metals and junctions based on them [5, 251]. To separate the contributions of different scattering mechanisms, a strong external magnetic field is used to influence the excitations having a nonzero magnetic moment.

1.12 Methods of Forming Point Contacts

Film structures containing point contacts are formed by vacuum deposition [142] (Figure 17). Initially, a primary metallic film is sputtered onto a dielectric substrate of glass, glass ceramic, crystalline quartz or sapphire. On the free surface of this film a dielectric coating is formed, the thickness and uniformity of which determines the length of the point contact and, to a significant degree, its diameter (usually $L \approx d$). The dielectric film is formed as a thin layer of native oxide or by deposition of some type of dielectric by vacuum sputtering. Its thickness is from a few nanometers up to a few tens of nanometers. Subsequent sputtering of a secondary film forms a metal-dielectric-metal tunnel structure with a high (10^5 - 10^7 ohm) resistance. Thicknesses of the metallic films should be substantially (on the order of one to two times) larger than the dimension of the point contact. The point contact is formed either by dielectric breakdown or brought about as a result of breaking the thin dielectric interlayer by keenly pressing on the free surface of the secondary film with a sharp needle. Dielectric breakdown and the subsequent "welding" of a point contact is accomplished by smoothly increasing the current until the voltage on the tunnel junction reaches the breakdown value.

The principal advantage of film structures is their high mechanical stability, which allows formation of point contacts with extremely small dimensions (from 1-2 nm.) and large resistances ($\sim 10^2$ ohms for ordinary s-p metals). A disadvantage of these contacts is that the polycrystalline structure of the film electrodes causes difficulties in studying the anisotropic EPI.

In a number of cases, shorted point contacts in film structures occur spontaneously (as a result of the intergrowth of metallic dendrites through defects in the dielectric layer or cracking of the latter during cooling or warming due to differences in the thermal expansion coefficients of the film structure elements). In this case, if the contacts satisfy the quality criteria described in section 1.13, they can be utilized in measurements.

Point contacts between bulk electrodes (including those between single crystals with given orientations) are usually formed by touching a sharp needle (with a radius of curvature from a few tenths up to a few tens of microns) to a planar electrode (*contacts of the needle-plane type*) [217].

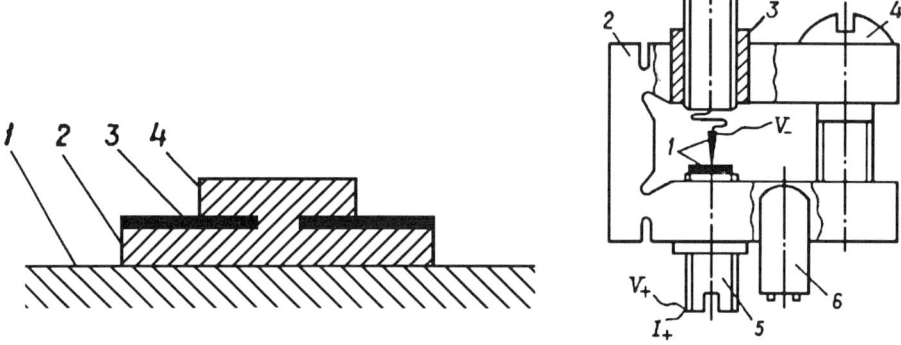

Figure 17 The cross section of a metal-dielectric-metal film structure containing a point contact:
   1 - substrate; 2 - primary metallic layer;
   3 - dielectric layer; 4 - secondary metallic layer

Figure 18 The arrangement for formation of needle-plane point contacts:
   1 - electrodes; 2 - hanger; 3 - Teflon sleeve; 4 - adjustment screw;
   5 - electrode connector; 6 - thermometer

The dimensions of a point contact and its quality depend on the applied force. This is regulated from outside the liquid helium cryostat in which the entire structure is placed (Figure 18). Before contact, the electrodes are usually mechanically polished. For removal of a defect layer from the surfaces of the electrodes, they are chemically (or electrochemically) etched and polished in a mixture of acids, the composition of which is specific for each metal, immediately before placing them into the pressing device. Reactants similar to those used for etching polished metal micro-sections are utilized with these acids [49, 88, 100].

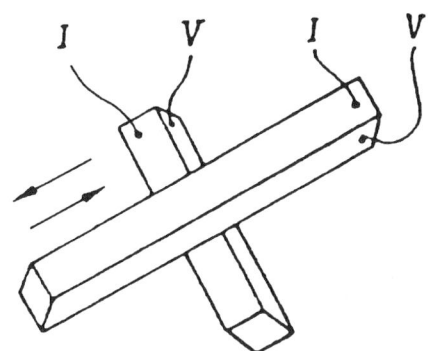

Figure 19
A sliding type contact ("edge-edge" geometry).

Prior to contact, the electrodes are usually covered by thin oxide layers which subsequently fulfill an important support function by providing mechanical stability to the contact. When the electrodes are pressed against one another, the greatest mechanical stress occurs at a point on the needle (or plane) where there is a micro-projection. Here, the oxide layer cracks, opening a small area with a direct metallic contact between the two electrodes.

A considerably more effective removal of microscopic residues (and plastic deformations of the metal) on the periphery of the contact occurs for what is called the sliding method of forming point contacts [122] (*point contacts of the sliding type*), in which the two electrodes are not only pressed against one another, but also slide on the contact surface, thus cleaning the contact region of contamination (Figure 19). In this case the metal in the region of the contact is deformed to a considerably smaller extent than for contacts of the needle-plane type.

Additional cleaning of impurities and defects from the metal may occur as a consequence of "welding" the pressed contacts with a current pulse [93].

1.13 Quality Criteria

In order to obtain reproducible point contact spectra it is necessary to select contacts suitable for investigation. The following criteria offer guidance in this regard.

1. The differential resistance of a contact increases with rising temperature or an increase in the applied potential bias (i.e. the contact has a metallic-type resistance). The indicated increase does not exceed a few tens of percent for a change in energy from 0 to $\hbar\omega_{max}$. In most cases the increase of the differential resistance in this energy interval is a few percent.

2. Large peaks in the second derivative of the current-voltage characteristics are observed in the fundamental energy region $0 \leq eV \leq \hbar\omega_{max}$ (usually the order of magnitude of $\hbar\omega_{max}$ is well known). For larger energies ($eV > \hbar\omega_{max}$) the magnitude of the second derivative of the current-voltage curve is nearly constant (background) (see Figure 16).

3. Transmission of electrons through the constriction corresponds to the ballistic regime (1). The various methods enumerated below permit estimation of the parameters entering into condition (1).

4. Features in the point contact spectra at small (with respect to $\hbar\omega_{max}$) bias, i.e. zero bias anomalies in the differential resistance, are absent.

5. The relative level of the background outside the limits of the EPI spectrum is not too large (say, $\gamma < 0.5$, see Figure 16).

The elastic mean free path, l_i and the diameter of the contact, d (in the circular aperture model), are determined from the temperature dependence of the contact resistance at zero bias in a region of sufficiently high temperature, where $R_0(T)$ is proportional to the resistivity $\rho(T)$ [3]:

$$d = \frac{d\rho/dT}{dR_0/dT} ,$$

$$l_i = \frac{\rho l_i}{d}\left[R_0 - \frac{16\rho l_i}{3\pi d^2}\right]^{-1}.$$

(25)

Here the product ρl_i is assumed to be well known for the given metal and independent of l_i.

The ratio $\nu = l_i/d$ is found indirectly from the following characteristics:

a) The intensity of the point contact spectrum $d^2I/dV^2(V)$ is approximately proportional to $\langle K(\nu) \rangle$. Therefore, if the intensity of a clean contact spectrum is known, ν is determined from the curve shown Figure 13 and then d and l_i are separately determined using equations (25) and the value R_0.

b) In the case when one or both of the electrodes can be (for a decrease in temperature) transferred into the superconducting state, the point contact can be related to the clean [63, 43] or dirty [63, 9] limit by the value of the critical current, I_c, and (or) excess current, I_{exc} (Table 3). In the latter case, the scattering centers are assumed to be uniformly distributed. For an inhomogeneous distribution of elastic scatterers of electrons (T-model, see Figure 2), the excess current decreases to zero with a decreasing transparency of the barrier. The current-voltage curve of a point contact thus approaches that of a tunnel junction [158, 159].

The ratio of the inelastic mean free path, l_ε, to the diameter of the aperture, d, determines the relative level of the background outside the limits of the EPI spectrum. For contacts conforming to the clean aperture model the following dependence of the background parameter on the resistance R_0 holds [140]:

$$\gamma \propto R_0^{-\frac{1}{2}}.$$

In the case where impurities and defects are forced to the periphery of the contact, forming a barrier to nonequilibrium phonons (see Figure 2b), the relative value of the background is directly proportional to the resistance of the contact R_0 [3].

The spectral density of low frequency ($f_0 = 10 - 10^6$ Hz) electrical fluctuations, S_V, can also be used for evaluation of the quality of a point contact. A clean point contact meeting the requirements of the ballistic regime corresponds to a nonmonotonic dependence of $S_V(eV)$, containing extrema in the regions of characteristic phonon frequencies [335].

Sharp extrema and other distinct features in the point contact spectra, which are reproduced at the same positions in energy for different contacts made of the same material, are evidence of long-range order in the atomic positions within the metal in the region of the constriction. Point contact spectra of metals containing a larger number of defects are characterized by broadened maxima.

Briefly, it is possible to formulate the following characteristics of EPI point contact spectra to serve as criteria for selecting high quality contacts: 1) *maximum intensity*, 2) *the presence of distinct features in the regions of phonon energies*, 3) *a minimal level of zero bias anomalies and background*.

Table 3
The critical and excess current of superconductor-superconductor point contacts with small dimensions ($d \ll \xi(T)$)

Limit	I_c (T = 0)	I_{exc}(T, V)
Clean ($l_i \gg d$)	$\pi \dfrac{\Delta_0}{eR_0}$	$\dfrac{8}{3} \dfrac{\Delta(T)}{eR_0} \tanh \dfrac{eV}{2kT}$
Dirty ($l_i \ll d$)	$1.32 \dfrac{\pi}{2} \dfrac{\Delta_0}{eR_0}$	$\left(\dfrac{\pi^2}{4} - 1 \right) \dfrac{\Delta(T)}{eR_0} \tanh \dfrac{eV}{2kT}$

Note: Δ_0 is the half-width of the energy gap for the superconductor at T = 0 and $\xi(T)$ is the coherence length in the superconductor.

1.14 Modulation Methods for Measuring Derivatives of the Current-Voltage Characteristics

If the current-voltage characteristic of a point contact is V(I) and the current I contains a small varying additive term in addition to a constant component I_0, that is

$$I = I_0 + i \cos 2\pi ft, \qquad i = \text{const},$$

then the voltage on the contact can be written as a Taylor series

$$V(I) = V_0 + v_1 \cos 2\pi ft + v_2 \cos 4\pi ft + v_3 \cos 6\pi ft + \ldots,$$

where

$$V_0 = V(I_0) + \frac{i^2}{4} \frac{d^2V}{dI^2} + \ldots,$$

$$v_1 = iR[1 + i^2(d^2V/dI^2)/8R + \ldots,$$

$$v_2 = \frac{i^2}{4} \frac{d^2V}{dI^2}[1 + i^2(d^4V/dI^4)\Big/48(d^2V/dI^2)] + \ldots,$$

$$v_3 = \frac{i^3}{24} \frac{d^3V}{dI^3} + \ldots,$$

and $R = dV/dI$ is the *differential resistance* of the contact.

Under the conditions

$$i^2 \ll \frac{8R}{d^3V/dI^3}, \qquad i^2 \ll \frac{48\, d^2V/dI^2}{d^4V/dI^4}$$

the first harmonic of the modulation signal $i\cos(2\pi ft)$ is proportional to the first derivative of the current-voltage characteristics, and the second harmonic to the second derivative:

$$v_1 = iR; \qquad v_2 = \frac{i^2}{4} \frac{d^2V}{dI^2}. \tag{26}$$

Let the current-voltage characteristics consist of smooth sections and features localized in the intervals ΔI and ΔV in current and voltage, respectively. Then equations (26) are correct if $i \ll \Delta I/2$ and $v_1 \ll \Delta V/2\pi$. In practice, the following designations are introduced for measurements of the effective values of v_1 and v_2

$$V_1 = \frac{i}{\sqrt{2}} \frac{dV}{dI}, \tag{27}$$

$$V_2 = \frac{i^2}{4\sqrt{2}} \frac{d^2V}{dI^2} \tag{28}$$

(V_1 is also called the modulation voltage).

The second derivative of the current-voltage characteristics, d^2V/dI^2, can be obtained using measurement schemes with a modulated signal source, operating as a constant current source:

$$r \gg R, \qquad i = \text{const.}, \tag{29}$$

where r is the internal resistance of the modulation signal source.

In some arrangements for measuring the second derivative of the current-voltage characteristics [114, 115] the amplitude of the current modulation automatically changes with changes in the differential resistance of the contact in such a way that the modulation voltage has a constant form. The operating regime of the modulation signal source in this case is characterized by the conditions

$$r \gg R, \qquad V_1 = \text{const.}, \tag{30}$$

and the effective value of the modulation signal voltage is proportional to the logarithmic derivative of the differential contact resistance:

$$S_2 = \frac{(V_1 = \text{const.})^2}{2\sqrt{2}} \frac{d}{dV} (\ln R).$$

For operation of the modulation signal in the constant voltage regime,

$$r \ll R, \qquad V_1 = \text{const.},$$

and the effective value of the voltage of the second harmonic of the modulation signal, $\widetilde{V}_2$, is proportional to the second derivative of the current-voltage characteristics, d^2I/dV^2.

Connections between the various second derivatives of the current-voltage characteristics are shown in Table 4. The second derivative of the current-voltage characteristics, dR/dV, appearing in Table 4, can be obtained using the measurement schemes suggested in reference [103] (also see the review article [74]).

The static resistance of the contact is

$$R_{st} = V/I = \frac{1}{I} \int_0^I R \, dI'.$$

Table 4
Conversion coefficients for different second derivatives of the current-voltage characteristics. (row = coefficient x column)

Derivative	$\dfrac{d^2I}{dV^2}$	$\dfrac{d(\ln R)}{dV}$	$\dfrac{dR}{dV}$	$\dfrac{d^2V}{dI^2}$
$\dfrac{d^2I}{dV^2}$	1	$-R^{-1}$	$-R^{-2}$	$-R^{-3}$
$\dfrac{d(\ln R)}{dV}$	$-R$	1	R^{-1}	R^{-2}
$\dfrac{dR}{dV}$	$-R^2$	R	1	R^{-1}
$\dfrac{d^2V}{dI^2}$	$-R^3$	R^2	R	1

Note: $R = dV/dI$, the differential resistance of the contact.

1.15 Modulation Broadening of Spectral Lines

For a finite amplitude of the modulation voltage, v_1, the second derivative of the current-voltage curve is recorded in the form [231]

$$F''(V) = \int_{-v_1}^{v_1} f''(V + v) \Phi(v) dv,$$

where V is the constant bias voltage, f ″(v) is the true second derivative of the current-voltage curve, and $\Phi(v)$ is the *modulation broadening function* for the spectral lines:

$$\Phi(v) = \frac{8}{3\pi} \frac{1}{v_1^4} (v_1^2 - v^2)^{\frac{3}{2}}.$$

The height of the bell-shaped curve $\Phi(v)$ is equal to $8/3\pi v_1$, and the width at half height is $1.22 v_1$. The latter quantity is taken to be the resolving power of the modulation method for measuring the second derivative of the current-voltage curve.

The resolution (in energy) of the point contact spectroscopy method is characterized by the quantity

$$\delta(eV) = [(5.44 kT)^2 + (1.22\sqrt{2} eV_1)^2]^{\frac{1}{2}},$$

equal to the width (at half maximum) of the bell-shaped curve, into which the extremely narrow lines of the point contact spectrum are transformed under temperature and modulation broadening. If the natural width at half maximum (that is, the width for T = 0, V_1 = 0) of a point contact spectrum peak is W_0, then the experimentally observed width is

$$W = [W_0^2 + \delta^2(eV)]^{\frac{1}{2}}.$$

1.16 Block Diagram of a Spectrometer

A block-diagram of the measurement apparatus for point contact spectroscopy is shown in Figure 20. Analogous schemes are utilized for tunneling spectroscopy of superconductors [253, 48], inelastic tunneling spectroscopy of thin dielectric films [31] and for investigation of semiconductor contacts. The point contact is placed either directly in liquid helium or in a thermostatically controlled chamber. The solenoid is used for quenching the superconductivity of specimens for analysis of superconducting metals below their critical temperature. The point contact is connected to the circuit by the four-probe method and is connected to a source of linearly variable bias, providing a variable direct current through the samples, within prescribed limits and at the necessary amplitude. (The basic schematic for such a device is described, for example, in reference [113].)

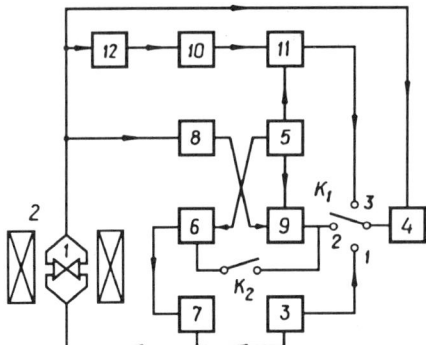

Figure 20 Block-diagram of a point contact spectrometer: 1 - point contact; 2 - solenoid; 3 - linear voltage source of the bias current; 4 - two-coordinate recording potentiometer; 5 - harmonic voltage generator; 6 - automatically regulated amplifier system; 7 - voltage-current transformer, 8, 9 - lock-in amplifier for detecting the first harmonic voltage; 10, 11 - lock-in amplifier for detecting the second harmonic voltage; 12 - filter

For measurements of the current-voltage characteristics, a voltage proportional to the current through the contact is recorded on a two-coordinate recording potentiometer as a function of the voltage bias on the contact (switch K_1 in Figure 20 in position 1).

For measurements of the voltage of the first or second harmonic of the modulation signal, a modulation current source is placed in parallel with the bias current source (switch K_1 corresponding to position 2 or 3). It consists of a harmonic voltage generator with little nonlinear distortion (< 0.02%), a system of automatic gain regulation, and a voltage-current converter. The modulation frequency, f, lies in the range from a few tens of hertz up to a few hundreds of kilohertz and is chosen to be a non-multiple of the power supply frequency. The voltage of the first or second harmonic of the modulation current signal originating in the point contact is fed into the Y-input of the recorder after detection and amplification. The bias voltage is fed into the X-input of the recorder. The voltage values of the first harmonic of the modulation signal are measured to an accuracy of 0.1% or better and serve as reference points for processing the point contact spectra.

If switch K_2 is open, the modulation signal source operates as a constant current source (29) and the voltage of the second harmonic is proportional to the derivative of the current-voltage characteristics, d^2V/dI^2. With switch K_2 closed, the modulation voltage source operates in regime (30) due to the negative feedback to the first harmonic modulation signal voltage and the voltage of the second harmonic is proportional to the derivative of the current-voltage characteristics $d(\ln R)/dV$.

A filter (Figure 21) [142] (which has a quality factor of ~ 100) is placed at the input of the lock-in amplifier for the second harmonic voltage and is used for suppression of the first harmonic signal and matching the low impedance of the contact to the high input resistance of the amplifier. The series C_2L, tuned to a resonance frequency f, is shunting the amplifier input at the modulation frequency. The circuit C_1C_2L is tuned to a resonance frequency of 2f and the second harmonic voltage is measured across the series C_2L (the impedance of which has a capacitive character at this frequency).

Suppression of the first harmonic signal at the input of the second harmonic voltage amplifier may be achieved by inserting the contact into a bridge circuit [131, 74, 32], with the sample in one arm of the bridge. The modulation signal is then applied on one diagonal of the bridge and the input of the second harmonic voltage amplifier is connected to the second diagonal of the bridge. The bridge circuit also allows small variations in the modulation signal, proportional to the first derivative of the current-voltage characteristics, to be recorded.

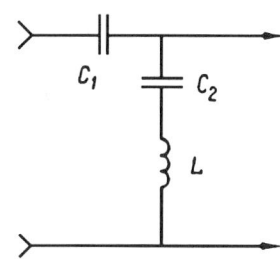

Figure 21
Filter schematic

1.17 The Procedure for Reconstruction of the Electron-phonon Interaction Point Contact Function from Measured Characteristics.

Reconstruction of the EPI point contact function from experimental data involves transformation of the second harmonic voltage $V_2(V) \propto d^2V/dI^2(V)$ or $S_2(V) \propto d(\ln R)/dV$ to the form $\tilde{V}_2(V) \propto d^2I/dV^2(V)$ (in this case a measurement circuit with a modulation signal source operating as a current source is used), calculation of the background from the dependence of $\tilde{V}_2(V)$ and normalization of the $g_{pc}(\omega)$ values obtained.

Due to the relationships between the second derivatives of the current-voltage characteristics (see Table 4), transformation of the point contact spectra to $\tilde{V}_2(V)$ reduces to division of $V_2(V)$ by the normalized variation of the modulation voltage to the third power or $S_2(V)$ by the normalized modulation voltage variation to first power. With this scaling the dependence of $V_2(V)$ or $S_2(V)$ is preserved in the function, $\tilde{V}_2(V)$. The scale (magnitude M_2) and, hence, the absolute value of the second harmonic voltage are determined directly by calibrating the measurement circuit or are found from the relationships

$$M_2 \approx \frac{1}{4\sqrt{2}}[V_1^2(V) - V_{1,0}^2] \Big/ \int_0^V V_2(V')\, dV', \qquad (31)$$

$$M_2 \approx \frac{\sqrt{2}}{4}[V_1(V) - (V_1 = const.)](V_1 = const.) \Big/ \int_0^V S_2(V')\, dV \qquad (32)$$

$$M_2 \approx \frac{\sqrt{2}}{4} \frac{[V_1(V) - (V_1 = const.)](V_1 = const.)^2}{V_1(V)} \Big/ \int_0^V \tilde{V}_2(V')\, dV'. \qquad (33)$$

If the scale of the second harmonic voltage is known, then from expressions (31) - (33) the function $V_1(V)$, proportional to the differential contact resistance, can be obtained from the voltage bias by means of identical transformations. Thus, to reconstruct the EPI point contact function curve, in addition to the point contact function proper in some kind of arbitrary units, it is necessary to know either two values of the differential resistance at different points in the spectrum or the value of the differential resistance at one point and the scale of the second harmonic voltage.

The background subtracted from $\tilde{V}_2(\omega)$ is chosen to be in the form of equation (24) and is calculated using an iteration method. The number of iterations is determined by the requirement of matching the function $g_{pc}(\omega)$ to a specified precision. As the zeroth approximation for the background it is convenient to use the function

$$B_0(V) = C_0 \int_0^V \tilde{V}_2(V')\,dV', \quad C_0 = \left.\frac{\tilde{V}_2(V)}{B_0(V)}\right|_{V=\hbar\omega_{max}/e}.$$

All known sources of deviation of real contacts from the clean aperture model lead only to a decrease in the absolute value of $g_{pc}(\omega)$ calculated from the measured contact characteristics using this model [337, 117]. Therefore, to determine the absolute value of the EPI point contact function for a specific metal the clean aperture model is used, and a point contact spectrum with the largest intensity is selected, provided that it is reproducible.

The EPI point contact function $g_{pc}(\omega)$ is determined by expressions which follow from formulas (2), (5), (27), (28) and the quantities given in Table 1, using the approximation of a quadratic isotropic electron dispersion relation:

$$g_{pc}(\omega) + B(eV) = -\frac{3\hbar v_F}{2\sqrt{2}e}\frac{\tilde{V}_2(eV)}{V_{1,0}^2 d} = -0.699\left(v_F/10^6\,\frac{m}{s}\right)\frac{\tilde{V}_2(eV)}{V_{1,0}^2 d},$$

$$d = \frac{4}{ek_F}\sqrt{\frac{\pi\hbar}{R_0}} = 45.5\,\frac{10^{10}\,m^{-1}}{k_F}\frac{1}{\sqrt{R_0}}.$$

Here, $\tilde{V}_2$ and $V_{1,0}$ are measured in volts, d is in nanometers, and R_0 is in ohms. The value of the Fermi wave vector $k_F = mv_F/\hbar$ is determined from the equation

$$k_F = \left(\frac{3\chi\pi^2}{\Omega}\right)^{\frac{1}{3}},$$

where χ is the number of valence electrons per unit cell and the volume of the unit cell is expressed through the lattice constants of the metal (Table 5).

Table 5
The unit cell volume for different crystal structures.

Crystal Lattice	Volume of a unit cell	Crystal Lattice	Volume of a unit cell
BCC	$a^3/2$	tetragonal	$a^2c/4$
FCC	$a^3/4$	rhombohedral	$abc/2$
hcp	$\frac{\sqrt{3}}{2}a^2c$		

The values of the lattice constants [107] and the corresponding values of the Fermi velocity indicated in Table 38 are useful in calculating the EPI point contact functions. In this same table, the values obtained for the EPI parameter, λ_{pc}, are compared with the values of the EPI parameter, λ, recommended in reference [201].

In the absence of data allowing one to determine the intensity of the point contact spectrum or the maximum intensity of spectra with sufficiently reproducible values, plots of $g_{pc}(\omega)$ are given in arbitrary units, with tables of the EPI point contact functions $g_{pc}(\omega)$ normalized to their maximum values.

1.18 Pseudopotential Calculations of the Electron-phonon Interaction Point Contact Functions

Calculation of the EPI point contact functions can be accomplished using pseudopotential theory. For these calculations, several approaches may be chosen. In the one, an ionic pseudopotential of the metal's crystal lattice is used twice: in a model describing the interaction between ions, the dynamic matrix of the crystal and the phonon spectrum are calculated, and then the pseudopotential is used for calculating the matrix elements of the EPI. In a second, the Born-Karmann method for finding the dynamic matrix [15] is used with results from inelastic neutron scattering and the pseudopotential is used only for the subsequent calculation of the

EPI function. As a rule, it is assumed that the Fermi surface is spherical and all anisotropy is attributable to anisotropy in the phonon spectrum [318]. In another, simpler, approach the phonon spectrum is assumed to be isotropic (Debye model), but a nonspherical model for the Fermi surface of the metal is used and features of the EPI point contact spectra generated by this factor are examined [36]. The most logical approach proposes to account for both the anisotropy of the phonon spectrum and the nonsphericity of the Fermi surface. Within the context of such an approach, calculations of the anisotropic EPI transport function of zinc have been carried out [315].

Detailed comparison of the calculated and experimentally obtained EPI point contact functions allows correction of the initial pseudopotential values used in the calculation. These values can be used for calculation of various characteristics of the metal: i.e., the phonon density of states, the thermodynamic EPI function $g(\omega)$, kinetic coefficients, etc..

1.19 Methods for Determining the Electron-phonon Interaction Function and the Phonon Density of States

McMillan and Rowell [252] proposed a procedure for determining the thermodynamic EPI function $g(\omega)$ of superconducting metals from experimental data of the tunneling density of states

$$N_T(\omega) = \sigma_S/\sigma_N = Re\left\{\omega/\sqrt{\omega^2 - \Delta^2(\omega)}\right\} \quad (34)$$

and the Eliashberg integral equations [293] at T = 0:

$$\Delta(\omega) = \frac{1}{Z(\omega)} \int_{\Delta_0}^{\omega_C} Re\left\{\frac{\Delta(\omega')}{\left[\omega'^2 - \Delta^2(\omega')\right]^{1/2}}\right\}\left[K^+(\omega, \omega') - \mu^*\right]d\omega',$$
$$(35)$$
$$\left[1 - Z(\omega)\right]\omega = \int_{\Delta_0}^{\infty} Re\left\{\frac{\omega'}{\left[\omega'^2 - \Delta^2(\omega')\right]^{1/2}}\right\}K^-(\omega, \omega')d\omega',$$

where

$$K^{\pm}(\omega, \omega') = \int_0^{\infty} g(\omega'')\left(\frac{1}{\omega'' + \omega' + \omega + i0^+} \pm \frac{1}{\omega'' + \omega' - \omega - i0^+}\right)d\omega''.$$

In expressions (34) and (35), σ_S and σ_N are the conductivity of a superconductor-insulator-normal metal tunneling contact in the superconducting and normal states, respectively, $\Delta_0 = \Delta(T = 0)$ is the half-width of the superconducting energy gap, $Z(\omega)$ is the renormalization function, μ^* is the coulombic pseudopotential, and ω_C is the cutoff frequency (usually $\omega_C \geq 5\omega_{max}$). Equations (35) are solved by iteration of the difference between experimental results and calculations from formula (34). The values of N_T and μ^* are used as fitting parameters. The McMillan-Rowell method is explained in detail, for example, in reference [21].

Recovery of the EPI function $g(\omega)$ from the Eliashberg equation is considerably simplified [23] by determining $\Delta(\omega)$ from the dispersion relation

$$Im\left\{\omega \Big/ \sqrt{\omega^2 - \Delta^2(\omega)}\right\} = \frac{2\omega}{\pi} \int_{\Delta_0}^{\infty} \frac{N_T(\omega) - N_{BCS}(\omega)}{\omega^2 - \omega'^2} d\omega'.$$

where

$$N_{BCS}(\omega) = Re\left\{\omega \Big/ \sqrt{\omega^2 - \Delta_0^2}\right\}$$

is the density of single-electron states in the Bardeen-Cooper-Schreiffer theory of superconductivity. The utilization of μ^* as a fitting parameter in this case is not required. The thermodynamic EPI function can also be found by utilizing the proximity tunneling effect [147, 330, 332].

Studies concerning the reconstruction of the phonon density of states in metals can conventionally be divided into three groups: calculations not utilizing experimental data, calculations modeled on the phonon dispersion curve measured by thermal neutron scattering or thermally diffused scattering of x-rays [15], and, finally, nonmodeled reconstructions of $F(\omega)$ in neutron experiments.

For the calculation of $F(\omega)$, the phonon frequencies and the polarization vector of lattice vibrations should be known for arbitrary phonon momenta. These quantities are generally determined experimentally only for a few (generally the principal) crystallographic directions. The Born-Karmann method is use for determination of the dispersion of phonons in arbitrary directions. Elements of the dynamic matrix of the crystal, coupling between themselves the lattice vibration frequencies and the polarization vectors, correspond to some simplified strong interaction model of crystal ions and are fit to the dispersion curve measured in the principle crystallographic directions. The phonon density of states calculated in that way contains distinct Van-Hove singularities and detailed fine structure. Errors in the determination of $F(\omega)$ depend on the nature of the approximation in the selected strong interaction model.

Neutron scattering studies using polycrystalline samples [17] allow recovery of F(ω) irrespective of the particular strong interaction model. However, the precision of such a method is not high and the resolution worsens appreciably with increasing phonon energies. The non-modeled function F(ω) is a broadened curve, drawn through experimental points, having a large spread and not containing sharp Van-Hove singularities. However, for vanadium, which has a coherent neutron scattering cross section small in comparison to the total cross section [30], such a method permits reconstruction of F(ω) to sufficiently high precision [101].

The phonon density of states can also be determined from precise measurements of the temperature dependence of the lattice specific heat [69]

$$C_V(T) = k \int_0^\infty B\left(\frac{\hbar\omega}{kT}\right) F(\omega)\, d\omega,$$

where

$$B(x) = \frac{x^2 e^x}{(e^x - 1)^2}.$$

However, reconstruction of F(ω) in this case is ascribed to the class of ill-posed problems, which are determined to be unstable relative to small changes of the input values. Therefore, acquisition of reliable information about F(ω) requires measurement of $C_V(T)$ to a precision presently unattainable in practice. Using special regularization methods, only a coarse representation of the function F(ω) can be generated [102]. Values of the EPI transport function $g_{tr}(\omega)$ reconstructed from precise measurements of the temperature dependence of the electrical resistance [214] are also ascribed to this class of problems.

References.
Theories of point contact spectroscopy: [9, 10, 36-38, 43, 50, 58-62, 64-68, 84-87, 108, 109, 126, 130, 148, 149, 170, 236, 244-248, 321, 323].

Point contact spectroscopy of alloys: [5, 8, 77, 79, 82, 106, 120, 129, 134, 164, 168, 178, 179, 190-192, 222, 239, 243].

Point contact spectroscopy of semiconductors: [46, 47, 127, 196, 215, 237, 271-273, 325].

Reviews of point contact spectroscopy: [16, pp. 388-403, 138, 140, 221]

Some reviews of the EPI: [41, 94, 201, 202, 253].

2. POINT CONTACT SPECTRA, ELECTRON-PHONON INTERACTION FUNCTIONS, AND THE PHONON DENSITY OF STATES IN METALS

2.1 Lithium

Crystal lattice: α-phase - BCC, β-phase - hcp

References. Point contact spectroscopy: [71, 80, 220, 244].
EPI functions: [209, 210, 244].
Phonon density of states: [151, 206, 235, 298].

Table 6 Values of the EPI point contact function for lithium.

$\hbar\omega$, meV	$g_{pc}(\omega)/[g_{pc}(\omega)]_{max}$ α-phase	$g_{pc}(\omega)/[g_{pc}(\omega)]_{max}$ β-phase	$\hbar\omega$, meV	$g_{pc}(\omega)/[g_{pc}(\omega)]_{max}$ α-phase	$g_{pc}(\omega)/[g_{pc}(\omega)]_{max}$ β-phase
0	0	0	22	0.485	0.380
1	0.000	0.000	23	0.473	0.321
2	0.000	0.013	24	0.469	0.272
3	0.011	0.026	25	0.453	0.226
4	0.022	0.053	26	0.439	0.191
5	0.080	0.086	27	0.421	0.149
6	0.189	0.154	28	0.399	0.121
7	0.378	0.233	29	0.370	0.103
8	0.551	0.316	30	0.342	0.090
9	0.686	0.393	31	0.320	0.077
10	0.792	0.494	32	0.307	0.077
11	0.860	0.637	33	0.305	0.075
12	0.931	0.687	34	0.297	0.072
13	0.993	0.802	35	0.273	0.070
14	0.992	0.892	36	0.238	0.075
15	0.958	0.966	37	0.177	0.079
16	0.867	0.999	38	0.102	0.059
17	0.739	0.951	39	0.029	0.040
18	0.638	0.848	40	0.000	0.029
19	0.535	0.694	41		0.011
20	0.509	0.564	42		0.000
21	0.498	0.452			

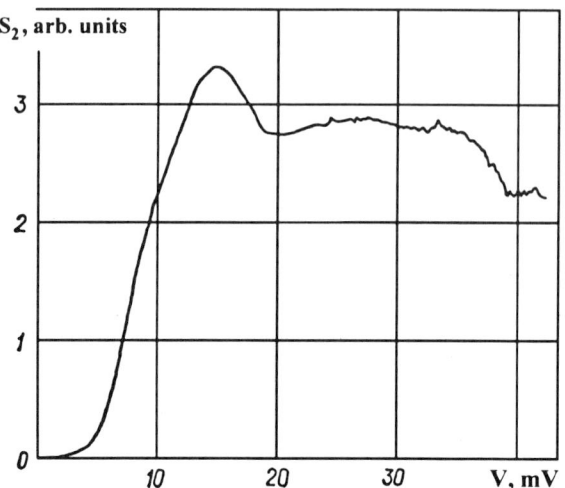

Figure 22 An EPI point contact spectrum for α-lithium [80] (a pressed point contact of the sliding type, R_0 = 15 ohms, V_1 = 700 μV, S_{2max} = 0.81 μV, T = 4.2 K).

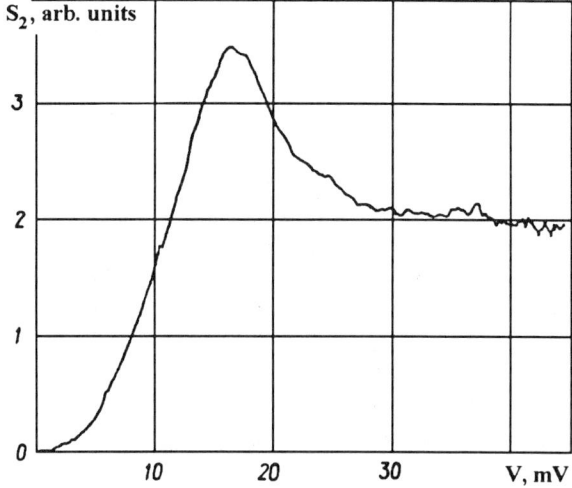

Figure 23 An EPI point contact spectrum for β-lithium [80] (a pressed point contact of the sliding type, R_0 = 6.5 ohms, V_1 = 800 μV, S_{2max} = 0.82 μV, T = 1.6 K).

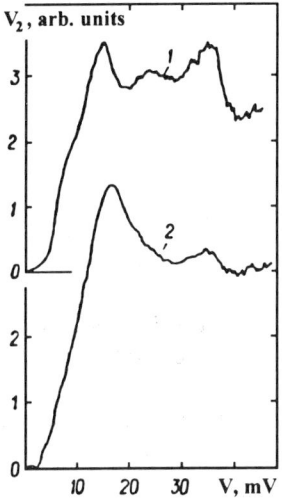

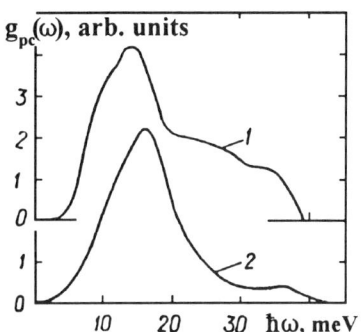

Figure 24 EPI point contact spectra for α-lithium (1) and β-lithium (2) [220] (a pressed point contact of a specialized construction, T = 1.5 K, R_0 = 2.5 ohms (1) and R_0 = 11.8 ohms (2)).

Figure 25 EPI point contact functions for lithium, reconstructed from the point contact spectra in Figures 22 and 23.
  1 - α-phase   2 - β-phase

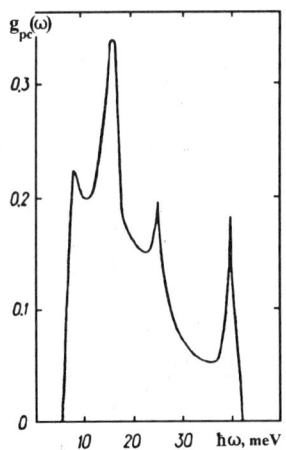

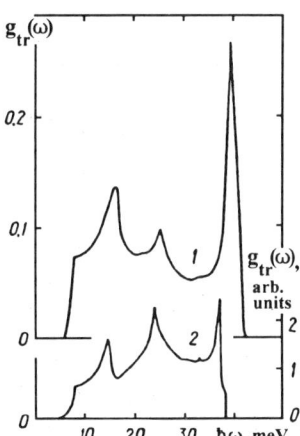

Figure 26 The EPI point contact functions for α-lithium (calculation [244], λ_{pc} = 0.665).

Figure 27 The EPI transport functions for α-lithium
  1, 2 - calculation [244] (λ_{tr} = 0.316), [210].

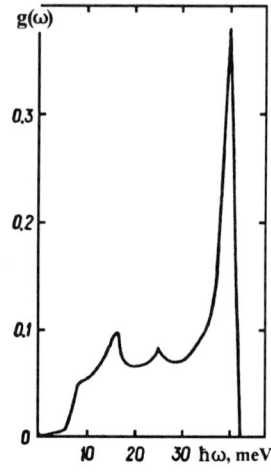

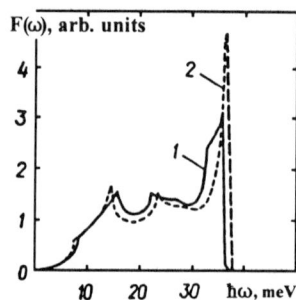

Figure 28 The thermodynamic EPI function for α-lithium (calculation [244], λ = 0.301).

Figure 29 The phonon density of states in α-lithium: 1 - calculation [151] with T = 293 K, 2 - calculation of Smith, et. al. with T = 98 K (cited in reference [151]).

2.2 Sodium

Crystal lattice: BCC

Table 7 Values of the EPI point contact function for sodium.

$\hbar\omega$, meV	$g_{pc}(\omega)$	$\hbar\omega$, meV	$g_{pc}(\omega)$	$\hbar\omega$, meV	$g_{pc}(\omega)$	$\hbar\omega$, meV	$g_{pc}(\omega)$
0	0	4.25	0.002	8.5	0.012	12.75	0.051
0.25	0.000	4.5	0.003	8.75	0.012	13	0.055
0.5	0.000	4.75	0.003	9	0.013	13.25	0.059
0.75	0.000	5	0.003	9.25	0.015	13.5	0.068
1	0.000	5.25	0.004	9.5	0.016	13.75	0.082
1.25	0.000	5.5	0.006	9.75	0.018	14	0.097
1.5	0.000	5.75	0.007	10	0.021	14.25	0.108
1.75	0.000	6	0.007	10.25	0.025	14.5	0.117
2	0.000	6.25	0.008	10.5	0.029	14.75	0.116
2.25	0.000	6.5	0.009	10.75	0.033	15	0.097
2.5	0.000	6.75	0.010	11	0.038	15.25	0.066
2.75	0.000	7	0.010	11.25	0.041	15.5	0.043
3	0.001	7.25	0.010	11.5	0.043	15.75	0.021
3.25	0.001	7.5	0.010	11.75	0.044	16	0.008
3.5	0.001	7.75	0.011	12	0.045	16.25	0.003
3.75	0.001	8	0.011	12.25	0.046	16.5	0.000
4	0.002	8.25	0.011	12.5	0.048		

References
Point contact spectroscopy: [37, 38, 71, 80, 130, 141, 145, 148, 220, 244].
EPI functions: [37, 38,1 30, 148, 165, 209, 211, 244].
Phonon density of states: [37, 38, 130, 156, 165, 181, 195, 200, 224, 235, 285, 298].

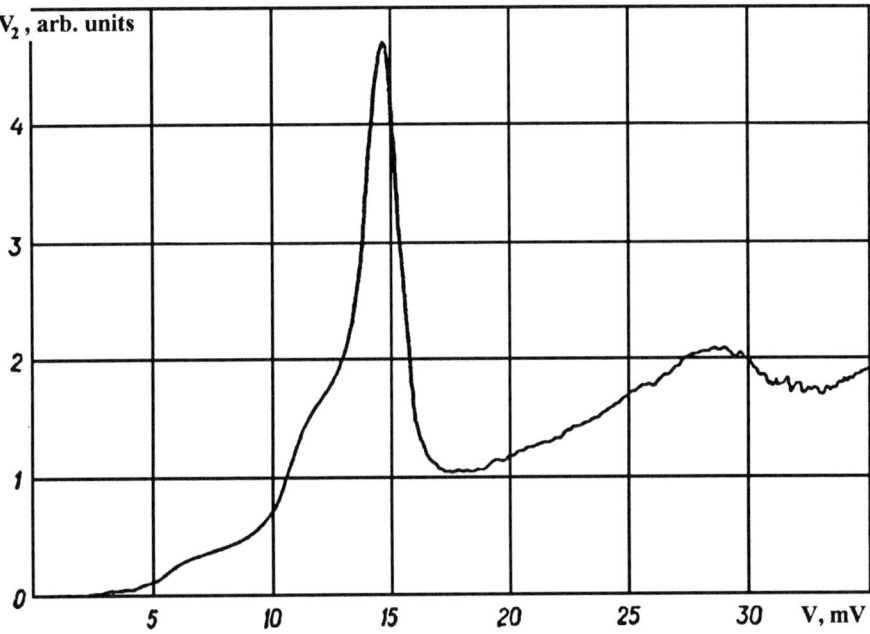

Figure 30 An EPI point contact spectrum for sodium [80] (a pressed point contact of the sliding type, $R_0 = 1$ ohm, $V_{1,0} = 280$ μV, $V_{2max} = 0.98$ μV, $T = 1.5$ K).

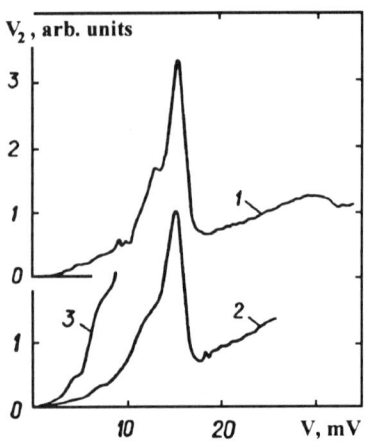

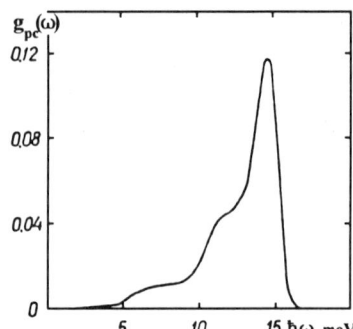

Figure 31 Point contact spectra for sodium according to the data of references [220] (1) and [145] (2,3):

1 - a pressed point contact of specialized construction, $R_0 = 1.1$ ohm, $T = 1.5$ K;
2 - a pressed point contact of the sliding type, $R_0 = 0.5$ ohm, $V_{1,0} = 400$ μV, $T = 1.7$ K;
3 - the initial segment of dependence 2 (with the scale of the ordinate axis expanded 5 times).

Figure 32 The EPI point contact functions for sodium, reconstructed from the point contact spectrum in Figure 30 ($\lambda_{pc} = 0.068$).

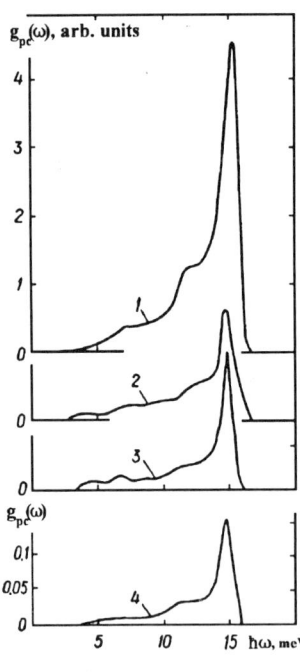

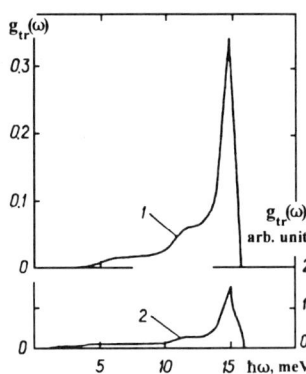

Figure 33 The EPI point contact function for sodium: 1, 2, 3, 4 - calculations [148], [38], [130], [244] ($\lambda_{pc} = 0.060$).

Figure 34 The EPI transport function for sodium: 1,2- calculation [244] ($\lambda_{tr} = 0.117$), [211].

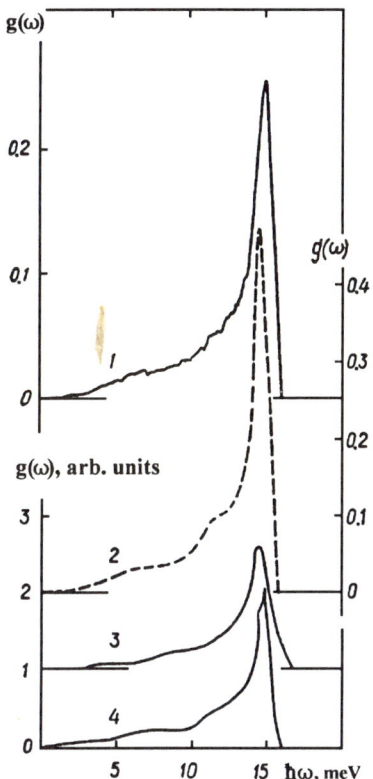 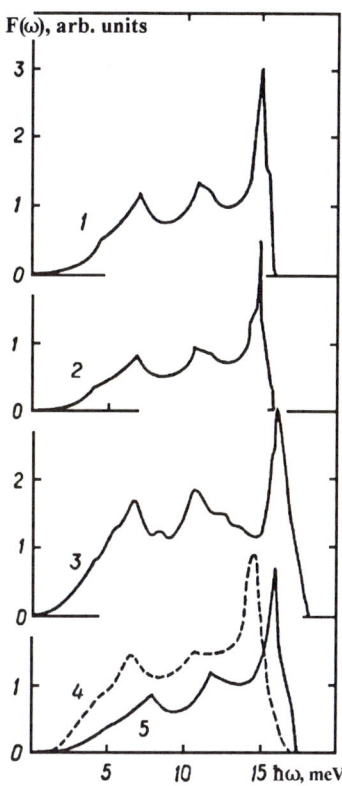

Figure 35 The EPI thermodynamic function for sodium: 1, 2, 3, 4 - calculations [148], [244], ($\lambda = 0.202$), [38], [130].

Figure 36 The phonon density of states in sodium : 1, 2, 3, 4, 5 - calculation [181], [200], [288], [38], [130].

2.3 Potassium

Crystal lattice: BCC
References. Point contact spectroscopy [37, 38, 80, 130, 141, 148, 220, 244, 248].
EPI functions: [37, 38, 130, 148, 165, 211, 244].
Phonon density of states: [37, 38, 130, 165, 177, 223, 235, 298].

Table 8 Values of the EPI point contact function for potassium.

$\hbar\omega$, meV	$g_{pc}(\omega)$	$\hbar\omega$, meV	$g_{pc}(\omega)$	$\hbar\omega$, meV	$g_{pc}(\omega)$
0	0	4	0.019	8	0.068
0.2	0.000	4.2	0.020	8.2	0.074
0.4	0.000	4.4	0.020	8.4	0.083
0.6	0.000	4.6	0.020	8.6	0.094
0.8	0.001	4.8	0.020	8.8	0.100
1	0.001	5	0.019	9	0.104
1.2	0.001	5.2	0.020	9.2	0.103
1.4	0.002	5.4	0.021	9.4	0.096
1.6	0.002	5.6	0.023	9.6	0.082
1.8	0.004	5.8	0.025	9.8	0.063
2	0.005	6	0.027	10	0.046
2.2	0.007	6.2	0.029	10.2	0.030
2.4	0.009	6.4	0.033	10.4	0.019
2.6	0.011	6.6	0.036	10.6	0.010
2.8	0.012	6.8	0.040	10.8	0.005
3	0.014	7	0.044	11	0.003
3.2	0.015	7.2	0.049	11.2	0.001
3.4	0.017	7.4	0.053	11.4	0.001
3.6	0.018	7.6	0.058	11.6	0.000
3.8	0.019	7.8	0.064		

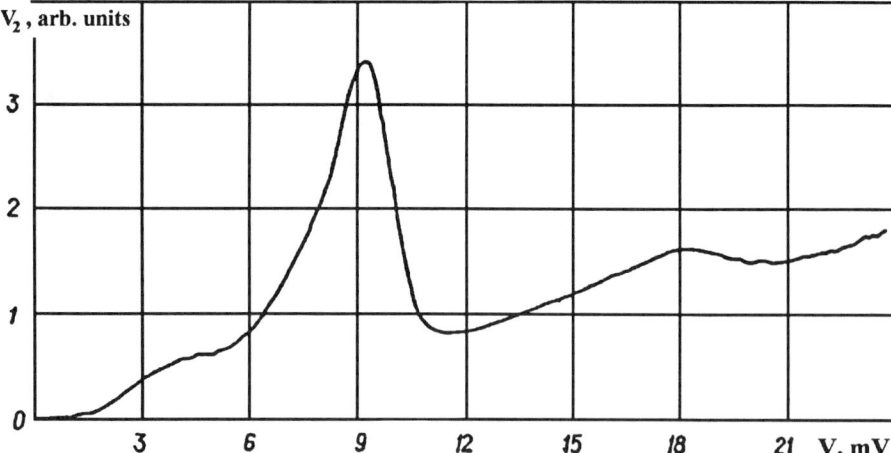

Figure 37 An EPI point contact spectrum for potassium [80] (a pressed point contact of the sliding type, R_0 = 2.4 ohms, $V_{1,0}$ = 350 μV, V_{2max} = 1.25 μV, T = 2.3 K).

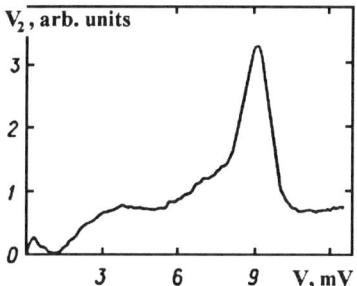
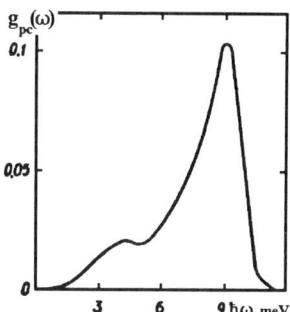

Figure 38 An EPI point contact spectrum for potassium [220] (a pressed point contact of a special construction, $R_0 = 2.9$ ohms, $T = 1.2$ K).

Figure 39 The EPI point contact function for potassium, reconstructed from the point contact spectrum in Figure 37 ($\lambda_{pc} = 0.11$).

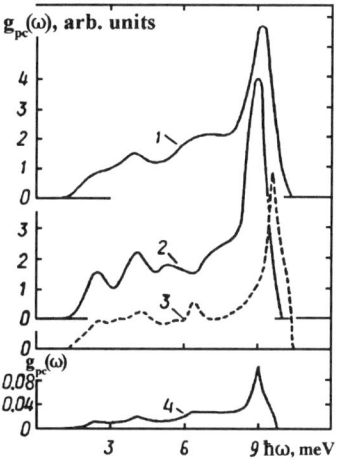

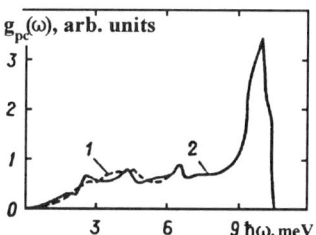

Figure 40 The EPI point contact function for potassium: 1, 2, 3, 4 - calculations [148], [38], [130], [244], ($\lambda_{pc} = 0.058$).

Figure 41 Anisotropy of the EPI point contact function for potassium (calculation [130]): 1 - [100] direction, 2 - [111]

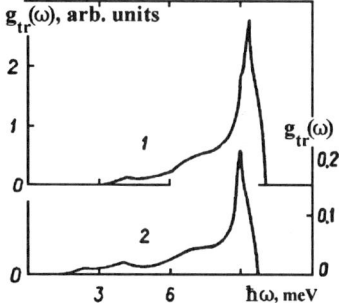

Figure 42 The EPI transport function for potassium:
1,2 - calculation [211],[244] ($\lambda_{tr} = 0.086$).

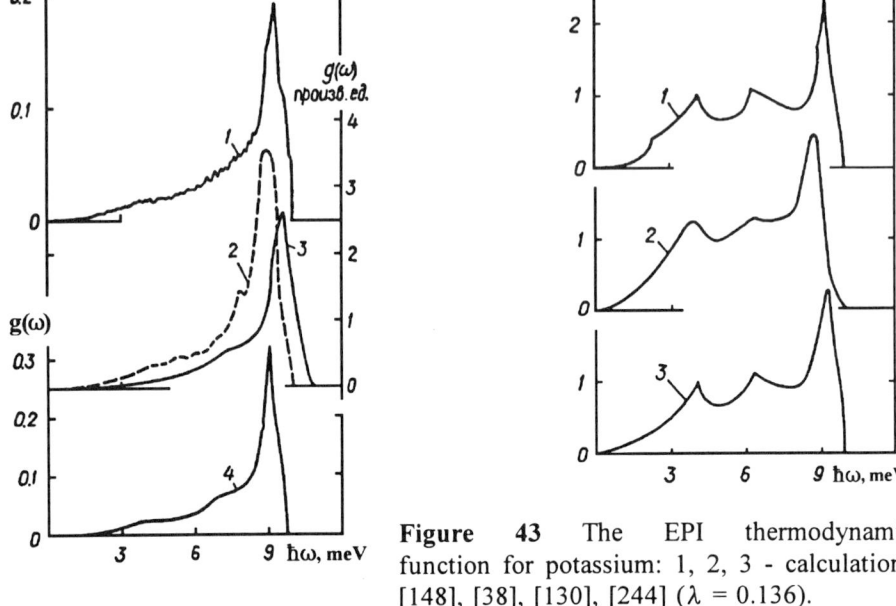

Figure 43 The EPI thermodynamic function for potassium: 1, 2, 3 - calculations [148], [38], [130], [244] ($\lambda = 0.136$).

Figure 44 The phonon density of states in potassium: 1, 2, 3 - calculations [177], [38], [130].

2.4 Copper

Crystal lattice: FCC

References. Point contact spectroscopy: [1, 3, 4, 13, 77, 133, 136, 142, 144, 145, 168, 217-219, 243, 312, 320, 322, 335, 337]. EPI functions: [150, 168, 225]. Phonon density of states: [27, 152, 188, 216, 242, 260, 262, 278, 299, 304, 306, 309, 313, 324].

Table 9 Values of the EPI point contact function for copper.

$\hbar\omega$, meV	$g_{pc}(\omega)$	$\hbar\omega$, meV	$g_{pc}(\omega)$	$\hbar\omega$, meV	$g_{pc}(\omega)$
0	0	2.5	0.002	5	0.003
0.5	0.001	3	0.002	5.5	0.003
1	0.001	3.5	0.002	6	0.003
1.5	0.001	4	0.002	6.5	0.004
2	0.001	4.5	0.003	7	0.004

Table 9, contd.

ℏω, meV	$g_{pc}(\omega)$	ℏω, meV	$g_{pc}(\omega)$	ℏω, meV	$g_{pc}(\omega)$
7.5	0.006	16	0.228	24.5	0.009
8	0.007	16.5	0.213	25	0.009
8.5	0.009	17	0.201	25.5	0.012
9	0.011	17.5	0.194	26	0.021
9.5	0.014	18	0.191	26.5	0.034
10	0.019	18.5	0.191	27	0.040
10.5	0.028	19	0.191	27.5	0.045
11	0.038	19.5	0.188	28	0.045
11.5	0.053	20	0.177	28.5	0.040
12	0.067	20.5	0.149	29	0.032
12.5	0.083	21	0.108	29.5	0.024
13	0.104	21.5	0.068	30	0.016
13.5	0.128	22	0.033	30.5	0.010
14	0.169	22.5	0.020	31	0.005
14.5	0.213	23	0.016	31.5	0.002
15	0.241	23.5	0.012	32	0.000
15.5	0.241	24	0.011		

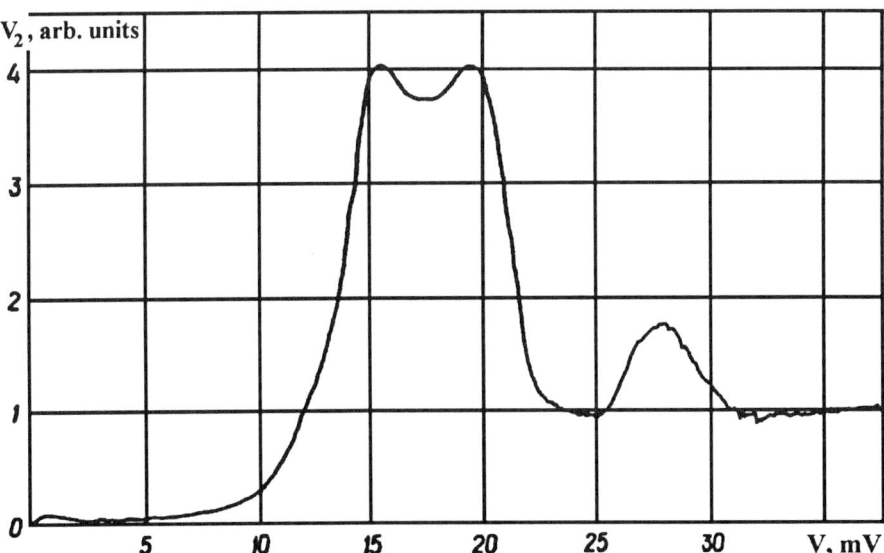

Figure 45 An EPI point contact spectrum for copper [335] (a pressed point contact of the sliding type, R_0 = 2.5 ohms, $V_{1,0}$ = 440 μV, V_{2max} = 1.09 μV, T = 1.6 K).

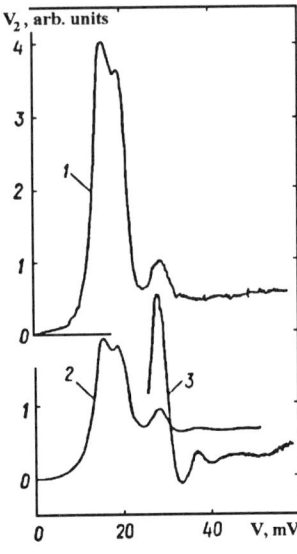

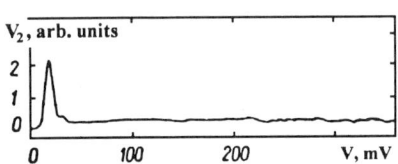

Figure 46 EPI point contact spectra for copper according to the data in references [221] (1) and [145] (2,3): 1 - needle-plane pressed contact, $R_0 = 3.3$ ohms, $T = 1.5$ K ; 2 - pressed contact of the sliding type, $R_0 = 3.1$ ohms, $V_{1,0} = 500$ μV, $T = 1.7$ K ; 3 -a portion of dependence 2 (with the scale of the ordinate axis expanded 8 times).

Figure 47 An EPI point contact spectrum for copper, measured to a large voltage [243] (needle-plane pressed contact, $R_0 = 7$ ohms, $V_{1,0} = 2050$ μV, $T = 4.2$ K).

Figure 48 Anisotropy of the EPI point contact spectra for copper [337] in the [100] (1), [110] (2), and [111] (3) directions (a needle-plane pressed contact):

1 - $R_0 = 2.75$ ohms
 $V_{1,0} = 200$ μV
 $V_{2max} = 0.25$ μV
 $T = 1.5$ K

2 - $R_0 = 10.5$ ohms
 $V_{1,0} = 400$ μV
 $V_{2max} = 0.21$ μV
 $T = 4.2$ K

3 - $R_0 = 4.45$ ohms
 $V_{1,0} = 300$ μV
 $V_{2max} = 0.21$ μV
 $T = 1.6$ K

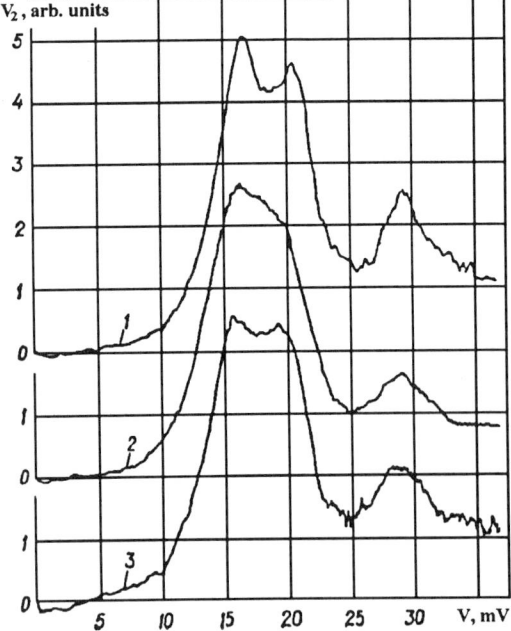

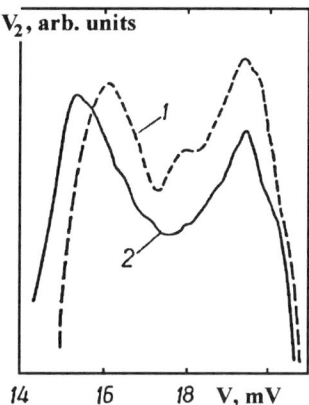
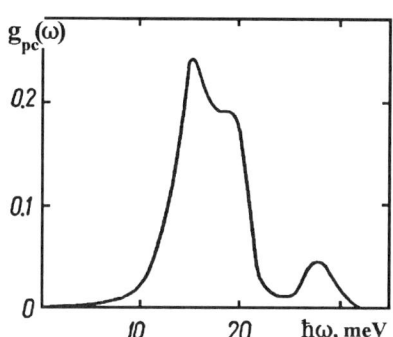

Figure 49 Anisotropy of the EPI point contact spectra for copper in the region of the TA maximum [337] : 1 - [110] direction ; 2 - [111]

Figure 50 The EPI point contact function for copper, reconstructed from the point contact spectrum in Figure 45 ($\lambda_{pc} = 0.24$).

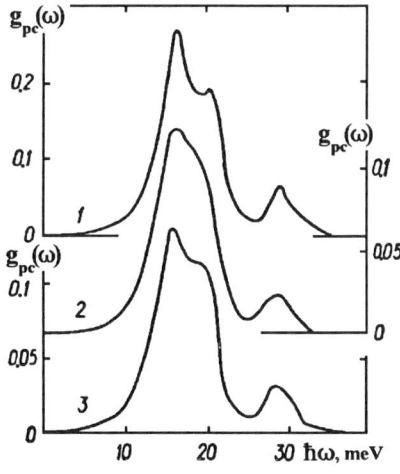

Figure 51 Anisotropy of the EPI point contact function for copper (plots reconstructed accordingly from the point contact spectra in Figure 48):
1 - [100] direction ; 2 - [110] ; 3 - [111]

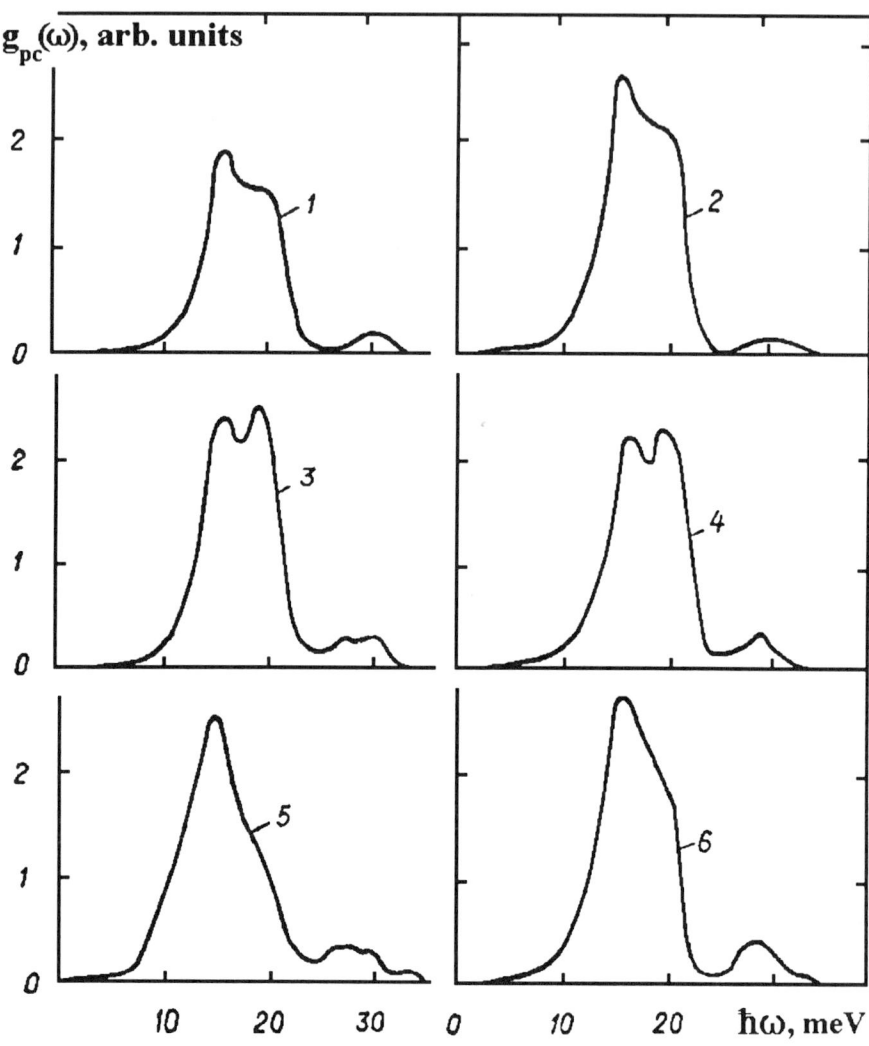

Figure 52 Anisotropy of the EPI point contact function for copper [168] in the [100] (1,2), [110] (3,4), and [111] (5,6) directions. The original point contact spectra were obtained in pressed needle-plane point contacts, $V_{1,0} = 850$ μV, $T = 1.5$ K with various R_0 values: 1 - $R_0 = 0.46$ ohms; 2 - 0.36 ohms; 3 - 21.6 ohms; 4 - 18.4 ohms; 5 - 2.9 ohms; 6 - 0.92 ohms.

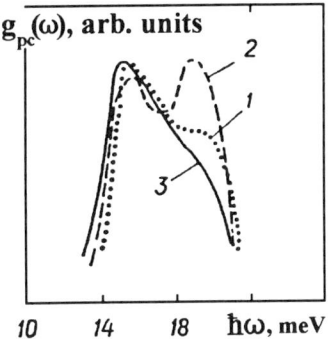

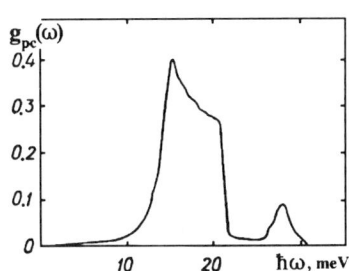

Figure 53 Anisotropy of the EPI point contact function for copper in the region of the TA maximum [168] :
1 - [100] direction; 2 - [110] ; 3 - [111].

Figure 54 The EPI point contact function for copper (calculation [168], $\lambda_{pc} = 0.34$).

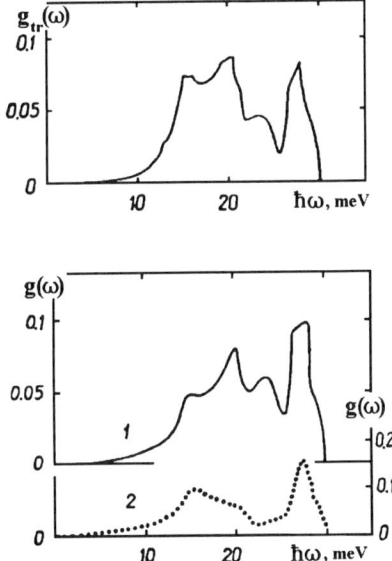

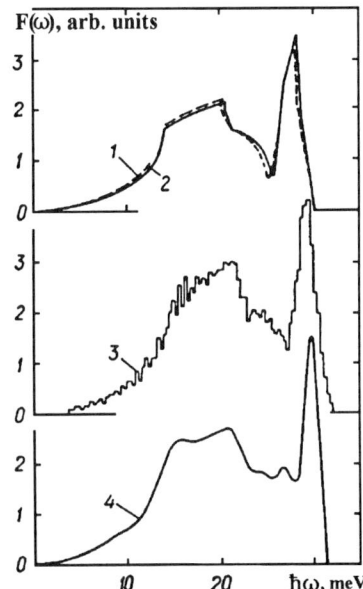

Figure 55 The EPI thermodynamic function for copper: (calculation [150], $\lambda_{tr} = 0.116$).

Figure 56 The EPI thermodynamic function for copper :
1,2 - calculation [150] ($\lambda = 0.111$), [168] ($\lambda = 0.14$).

Figure 57 The phonon density of states in copper: 1,2 - calculation [242] (1 - T = 298 K, 2 - 49 K); 3,4 - calculation [188], [313].

2.5 Silver

Crystal lattice : FCC
References. Point contact spectroscopy : [8, 124, 141, 142, 217-219, 335].
Phonon density of states : [152, 155, 205, 242].

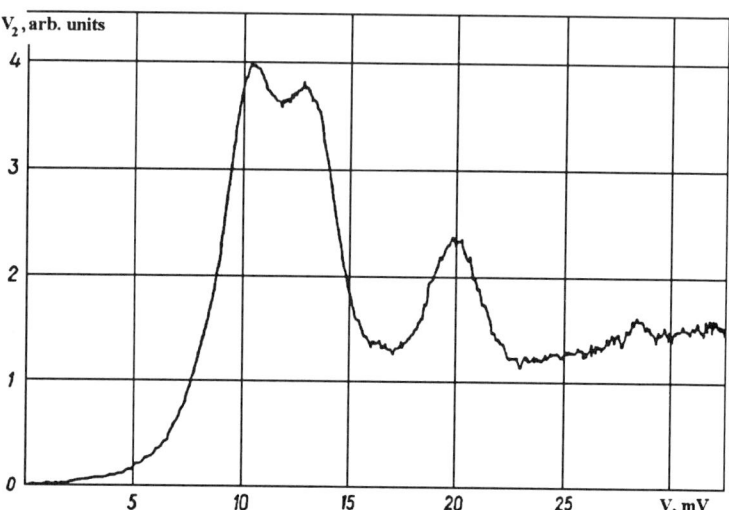

Figure 58 An EPI point contact spectrum for silver [81] (a pressed needle-plane point contact, $R_0 = 0.95$ ohms, $V_{1,0} = 450$ μV, $V_{2max} = 0.88$ μV, T = 1.6 K).

Table 10 Values of the EPI point contact function for silver.

ℏω, meV	$g_{pc}(\omega)$	ℏω, meV	$g_{pc}(\omega)$	ℏω, meV	$g_{pc}(\omega)$
0	0	4.25	0.002	8.5	0.040
0.25	0.000	4.5	0.002	8.75	0.046
0.5	0.000	4.75	0.002	9	0.053
0.75	0.000	5	0.004	9.25	0.061
1	0.000	5.25	0.005	9.5	0.071
1.25	0.000	5.5	0.006	9.75	0.079
1.5	0.000	5.75	0.006	10	0.086
1.75	0.000	6	0.007	10.25	0.089
2	0.000	6.25	0.008	10.5	0.089
2.25	0.001	6.5	0.010	10.75	0.087
2.5	0.001	6.75	0.012	11	0.083
2.75	0.001	7	0.014	11.25	0.079
3	0.001	7.25	0.017	11.5	0.075
3.25	0.002	7.5	0.021	11.75	0.072
3.5	0.002	7.75	0.025	12	0.071
3.75	0.002	8	0.029	12.25	0.071
4	0.002	8.25	0.034	12.5	0.071

Table 10, contd.

ℏω, meV	$g_{pc}(\omega)$	ℏω, meV	$g_{pc}(\omega)$	ℏω, meV	$g_{pc}(\omega)$
12.75	0.071	16.25	0.008	19.75	0.026
13	0.070	16.5	0.007	20	0.026
13.25	0.067	16.75	0.006	20.25	0.025
13.5	0.063	17	0.006	20.5	0.022
13.75	0.057	17.25	0.006	20.75	0.019
14	0.050	17.5	0.007	21	0.015
14.25	0.042	17.75	0.007	21.25	0.012
14.5	0.034	18	0.009	21.5	0.009
14.75	0.026	18.25	0.011	21.75	0.006
15	0.019	18.5	0.013	22	0.004
15.25	0.015	18.75	0.017	22.25	0.002
15.5	0.012	19	0.020	22.5	0.001
15.75	0.010	19.25	0.023	22.75	0.000
16	0.009	19.5	0.025		

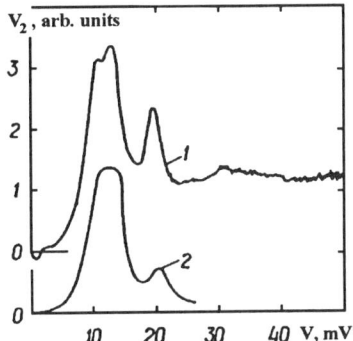
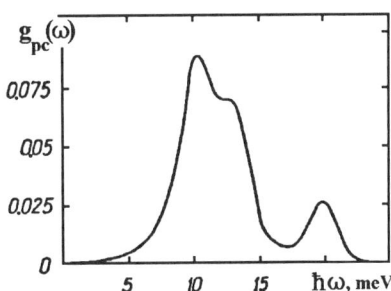

Figure 59 EPI point contact spectra for silver from the data in references [81] (1) and [218] (3) (a pressed needle-plane point contact):
1 - R_0 = 90 ohms, $V_{1,0}$ = 700 μV, T = 1.6 K ;
2 - R_0 = 16.3 ohms, $V_{1,0}$ = 300 μV, T = 1.2 K.

Figure 60 The EPI point contact function for silver, reconstructed from the point contact spectrum in Figure 58 (λ_{pc} = 0.1).

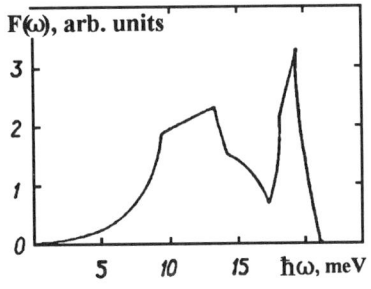

Figure 61 The phonon density of states in silver (calculation [242]).

2.6 Gold

Crystal lattice: FCC
References. Point contact spectroscopy: [81, 124, 141, 217-219, 335].
EPI functions: [214]. Phonon density of states: [52, 152, 205, 242, 275].

Table 11 Values of the EPI point contact function for gold.

$\hbar\omega$, meV	$g_{pc}(\omega)$	$\hbar\omega$, meV	$g_{pc}(\omega)$	$\hbar\omega$, meV	$g_{pc}(\omega)$
0	0	7	0.029	14	0.004
0.25	0.000	7.25	0.035	14.25	0.003
0.5	0.000	7.5	0.042	14.5	0.002
0.75	0.000	7.75	0.051	14.75	0.002
1	0.000	8	0.061	15	0.002
1.25	0.000	8.25	0.068	15.25	0.002
1.5	0.000	8.5	0.074	15.5	0.004
1.75	0.000	8.75	0.079	15.75	0.005
2	0.000	9	0.083	16	0.007
2.25	0.001	9.25	0.084	16.25	0.010
2.5	0.001	9.5	0.086	16.5	0.014
2.75	0.001	9.75	0.088	16.75	0.017
3	0.001	10	0.089	17	0.019
3.25	0.001	10.25	0.089	17.25	0.020
3.5	0.002	10.5	0.085	17.5	0.021
3.75	0.002	10.75	0.078	17.75	0.020
4	0.003	11	0.069	18	0.019
4.25	0.004	11.25	0.057	18.25	0.017
4.5	0.005	11.5	0.044	18.5	0.014
4.75	0.006	11.75	0.032	18.75	0.012
5	0.007	12	0.022	19	0.010
5.25	0.009	12.25	0.016	19.25	0.007
5.5	0.011	12.5	0.012	19.5	0.005
5.75	0.013	12.75	0.010	19.75	0.004
6	0.015	13	0.008	20	0.003
6.25	0.018	13.25	0.006	20.25	0.002
6.5	0.021	13.5	0.005	20.5	0.001
6.75	0.025	13.75	0.004	20.75	0.000

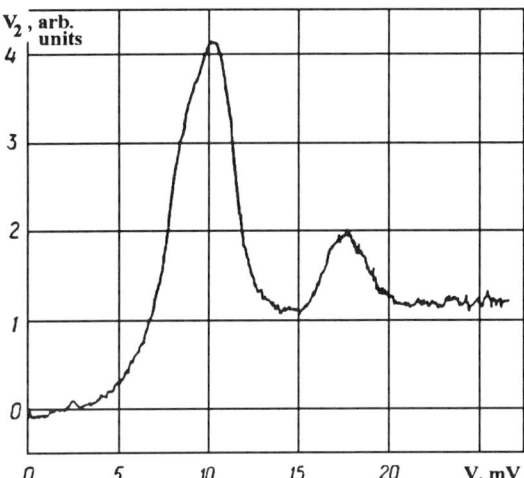

Figure 62 An EPI point contact spectrum for gold [81] (a pressed needle-plane point contact, $R_0 = 1.44$ ohms, $V_{1,0} = 318$ μV, $V_{2\,max} = 0.32$ μV, T = 1.6 K).

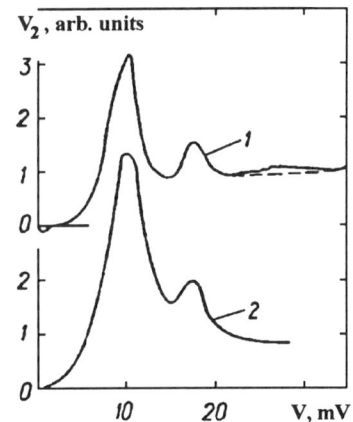
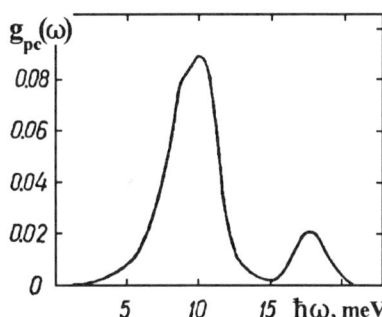

Figure 63 EPI point contact spectra for gold :
1 - data of reference [81] (the same point contact as that of Figure 62, $V_{1,0} = 45$ μV);
2 - data of reference [218] (a pressed needle-plane point contact, $R_0 = 32$ ohms, $V_{1,0} = 300$ μV, T = 1.2 K).

Figure 64 The EPI point contact function for gold, reconstructed from the point contact spectrum in Figure 62 ($\lambda_{pc} = 0.095$).

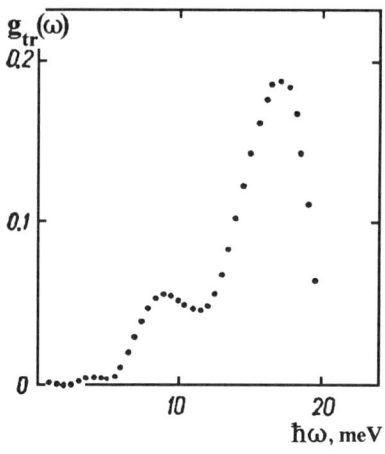
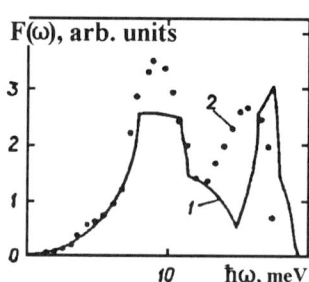

Figure 65 The EPI transport function in gold (calculation [214]).

Figure 66 The phonon density of states in gold: 1, 2 - calculation [242], [275].

2.7 Beryllium

Crystal lattice: hcp
References. Point contact spectroscopy: [77, 78, 141].
Phonon density of states: [19, 99, 160, 254, 284, 316, 339].

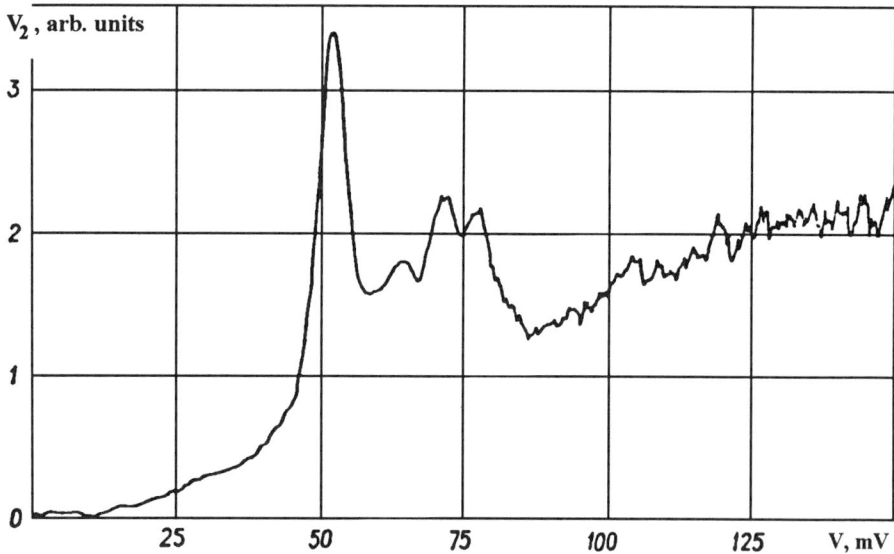

Figure 67 An EPI point contact spectrum for beryllium [78] (a pressed point contact of the sliding type, $R_0 = 8.5$ ohms, $V_{1,0} = 1050$ μV, $V_{2max} = 2.76$ μV, $T = 4.2$ K).

Table 12 Values of the EPI point contact function for beryllium.

ℏω, meV	$g_{pc}(\omega)$	ℏω, meV	$g_{pc}(\omega)$	ℏω, meV	$g_{pc}(\omega)$
0	0	30	0.027	60	0.099
1	0.000	31	0.028	61	0.100
2	0.000	32	0.029	62	0.105
3	0.000	33	0.030	63	0.109
4	0.001	34	0.031	64	0.108
5	0.001	35	0.033	65	0.103
6	0.002	36	0.035	66	0.094
7	0.002	37	0.037	67	0.082
8	0.002	38	0.041	68	0.089
9	0.002	39	0.044	69	0.109
10	0.002	40	0.049	70	0.120
11	0.002	41	0.053	71	0.130
12	0.003	42	0.057	72	0.128
13	0.004	43	0.062	73	0.112
14	0.005	44	0.068	74	0.095
15	0.006	45	0.078	75	0.091
16	0.007	46	0.099	76	0.097
17	0.008	47	0.127	77	0.100
18	0.009	48	0.176	78	0.093
19	0.010	49	0.235	79	0.075
20	0.012	50	0.291	80	0.054
21	0.014	51	0.359	81	0.040
22	0.014	52	0.375	82	0.031
23	0.015	53	0.330	83	0.023
24	0.018	54	0.256	84	0.016
25	0.018	55	0.186	85	0.009
26	0.020	56	0.139	86	0.003
27	0.022	57	0.113	87	0.000
28	0.024	58	0.103		
29	0.025	59	0.100		

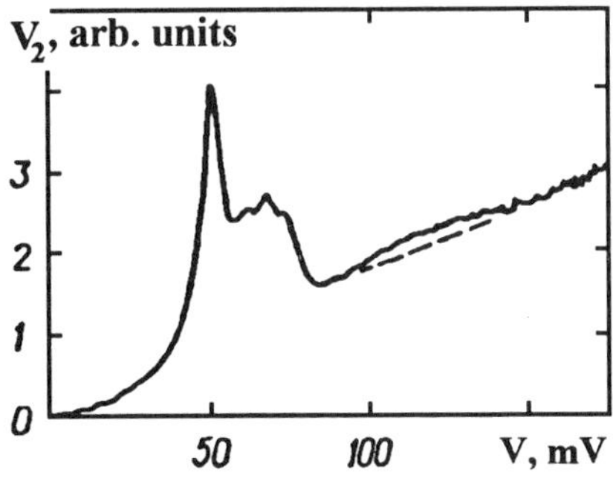

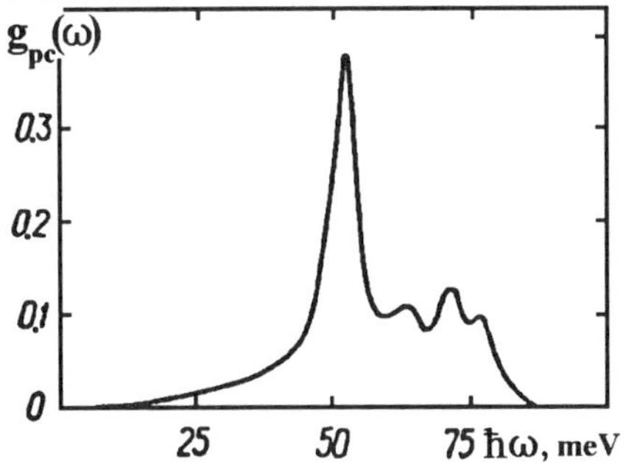

Figure 68 An EPI point contact spectrum for beryllium [78] (a pressed point contact of the sliding type, $R_0 = 7$ ohms, $V_{1,0} = 1340$ μV, T = 4.2 K).

Figure 69 The EPI point contact function for beryllium, reconstructed from the point contact spectrum in Figure 67 ($\lambda_{pc} = 0.24$).

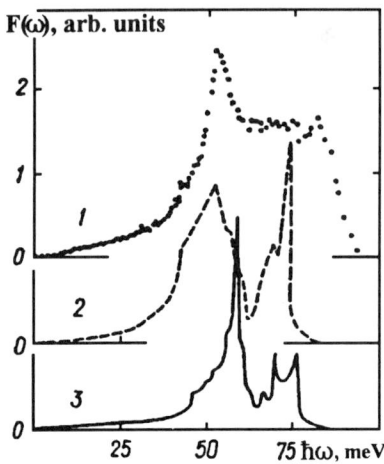

Figure 70 The phonon density of states in beryllium:
1 - experimental data [19], obtained by thermal neutron scattering;
2, 3 - calculations [339], [284].

2.8 Magnesium

Crystal lattice: hcp
References. Point contact spectroscopy: [83, 141]. EPI functions: [162].
Phonon density of states: [20, 33, 34, 55, 160, 186, 279, 284, 285, 316, 339].

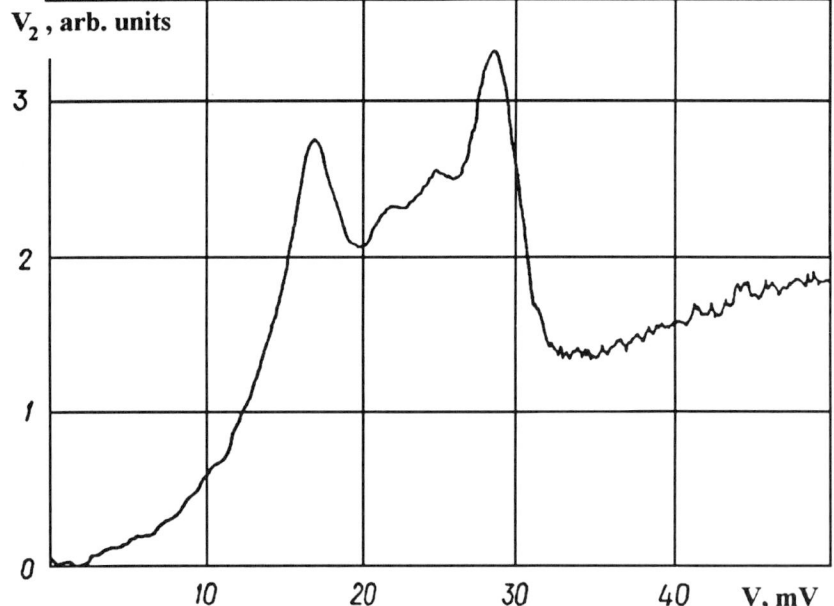

Figure 71 An EPI point contact spectrum for magnesium [83] (a pressed needle-plane point contact, $R_0 = 11$ ohms, $V_{1,0} = 500$ μV, $V_{2max} = 0.69$ μV, $T = 1.7$ K).

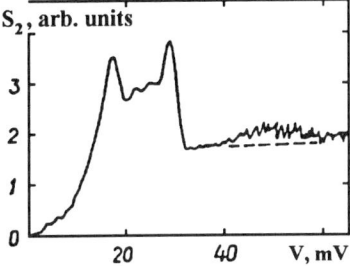
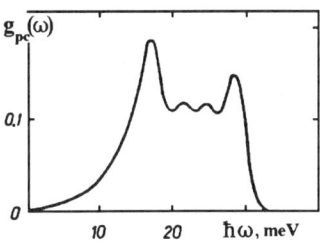

Figure 72 An EPI point contact spectrum for magnesium [83] (the same contact as that of Figure 71, $V_1 = 600$ μV).

Figure 73 The EPI point contact function for magnesium, reconstructed from the point contact spectrum in Figure 71 ($\lambda_{pc} = 0.28$).

Table 13 Values of the EPI point contact function for magnesium.

ℏω, meV	$g_{pc}(\omega)$	ℏω, meV	$g_{pc}(\omega)$	ℏω, meV	$g_{pc}(\omega)$
0	0	11.5	0.054	23	0.109
0.5	0.001	12	0.061	23.5	0.109
1	0.001	12.5	0.069	24	0.114
1.5	0.002	13	0.077	24.5	0.117
2	0.002	13.5	0.089	25	0.116
2.5	0.004	14	0.100	25.5	0.111
3	0.004	14.5	0.115	26	0.106
3.5	0.006	15	0.124	26.5	0.109
4	0.007	15.5	0.142	27	0.117
4.5	0.008	16	0.165	27.5	0.131
5	0.008	16.5	0.180	28	0.145
5.5	0.009	17	0.187	28.5	0.149
6	0.012	17.5	0.177	29	0.139
6.5	0.014	18	0.156	29.5	0.116
7	0.016	18.5	0.138	30	0.089
7.5	0.018	19	0.118	30.5	0.057
8	0.021	19.5	0.110	31	0.033
8.5	0.025	20	0.108	31.5	0.018
9	0.028	20.5	0.110	32	0.007
9.5	0.033	21	0.115	32.5	0.003
10	0.037	21.5	0.118	33	0.001
10.5	0.042	22	0.116	33.5	0.001
11	0.046	22.5	0.114	34	0.000

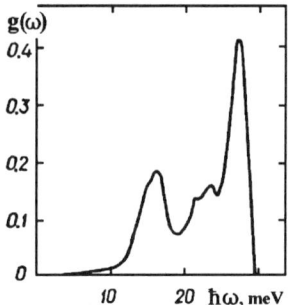

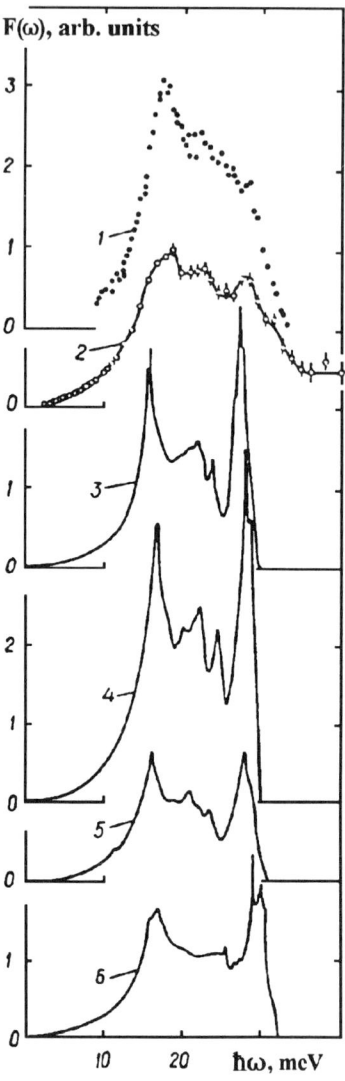

Figure 74 The EPI thermodynamic function for magnesium [162], obtained by the proximity tunnel effect method ($\lambda = 0.29 \pm 0.03$).

Figure 75 The phonon density of states in magnesium:
1, 2 - experimental data [20], [33], obtained by thermal neutron scattering ;
3, 4, 5, 6 - calculations [279], [186], [285], [284].

2.9 Zinc

Crystal lattice: hcp
References. Point contact spectroscopy: [12, 13, 32, 132, 133, 139]. EPI functions [314, 315, 318]. Phonon density of states: [33, 55, 160, 172, 254, 284, 316, 318, 339].

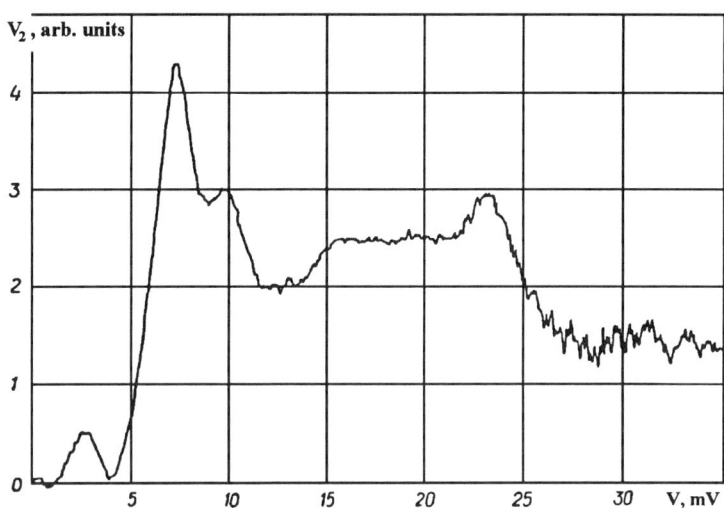

Figure 76 EPI point contact spectrum for zinc [139] (point contact obtained by the thin film technique, $R_0 = 29.1$ ohms, $V_{1,0} = 350$ μV, $V_{2max} = 0.2$ μV, T = 1.5 K).

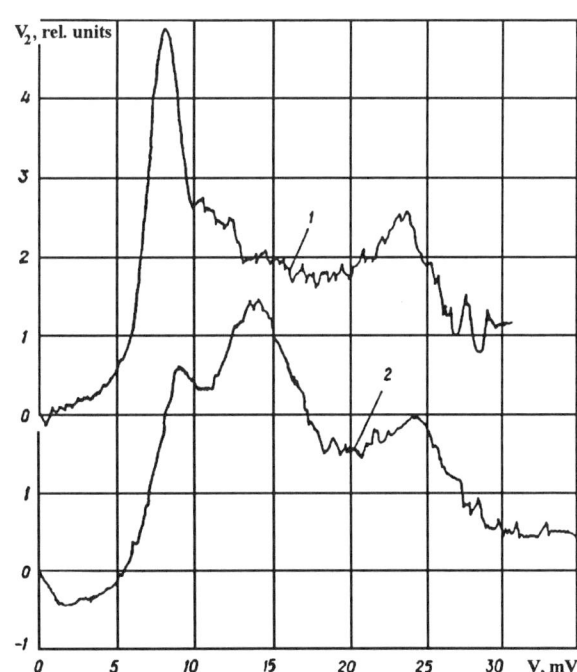

Figure 77 Anisotropy of the EPI point contact spectra for zinc [132] in the [0001] (1) and [11$\bar{2}$0] (2) directions (pressed needle-plane point contacts, T = 1.5 K) :

1 - $R_0 = 9.7$ ohms,
$V_{1,0} = 300$ μV,
$V_{2max} = 0.14$ μV

2 - $R_0 = 9.8$ ohms,
$V_{1,0} = 500$ μV,
$V_{2max} = 0.11$ μV

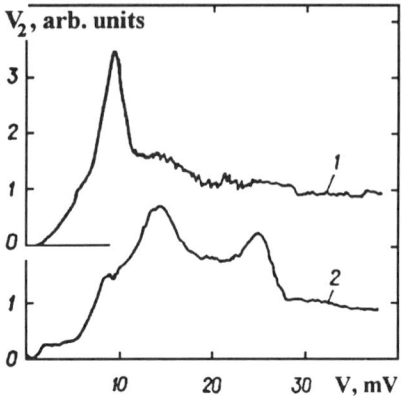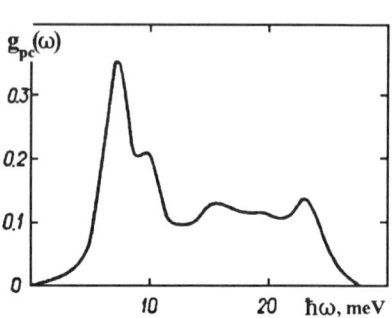

Figure 78 EPI point contact spectra for zinc [32] (a pressed needle-plane point contact, $V_{1,0} = 200$ μV, T = 1.5 K) :
1 - [0001] direction,   2 - (0001) plane.

Figure 79 The EPI point contact function for zinc, reconstructed from the point contact spectrum in Figure 76 ($\lambda_{pc} = 0.57$).

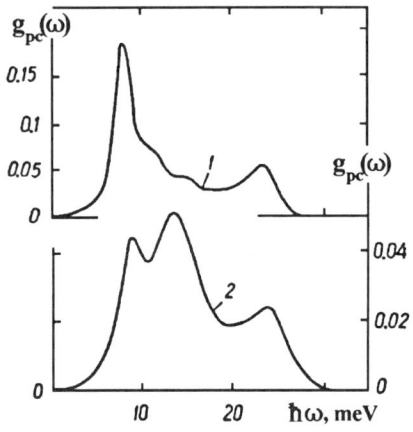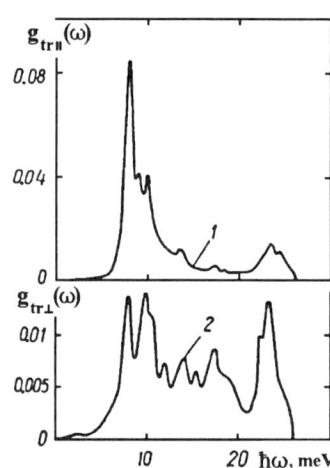

Figure 80 Anisotropy of the EPI point contact function for zinc (plots reconstructed from the corresponding point contact spectra in Figure 77): 1 - [0001] direction ; 2 - [11$\bar{2}$0].

Figure 81 The EPI transport function for zinc, calculated in [315] for directions parallel to [0001] (1) and perpendicular to [0001] (2) in the axially symmetric model.

Table 14 Values of the EPI point contact function for zinc.

$\hbar\omega$, meV	$g_{pc}(\omega)$	$\hbar\omega$, meV	$g_{pc}(\omega)$	$\hbar\omega$, meV	$g_{pc}(\omega)$
0	0	9.5	0.206	19	0.114
0.25	0.000	9.75	0.207	19.25	0.114
0.5	0.001	10	0.203	19.5	0.114
0.75	0.001	10.25	0.191	19.75	0.113
1	0.001	10.5	0.173	20	0.112
1.25	0.002	10.75	0.156	20.25	0.110
1.5	0.004	11	0.139	20.5	0.108
1.75	0.005	11.25	0.122	20.75	0.106
2	0.006	11.5	0.108	21	0.105
2.25	0.008	11.75	0.101	21.25	0.105
2.5	0.009	12	0.099	21.5	0.106
2.75	0.011	12.25	0.097	21.75	0.108
3	0.013	12.5	0.096	22	0.110
3.25	0.017	12.75	0.096	22.25	0.115
3.5	0.021	13	0.095	22.5	0.121
3.75	0.025	13.25	0.097	22.75	0.129
4	0.029	13.5	0.098	23	0.134
4.25	0.034	13.75	0.099	23.25	0.135
4.5	0.039	14	0.102	23.5	0.131
4.75	0.044	14.25	0.107	23.75	0.123
5	0.056	14.5	0.112	24	0.112
5.25	0.081	14.75	0.118	24.25	0.100
5.5	0.113	15	0.122	24.5	0.086
5.75	0.145	15.25	0.125	24.75	0.072
6	0.181	15.5	0.128	25	0.059
6.25	0.221	15.75	0.127	25.25	0.049
6.5	0.264	16	0.126	25.5	0.042
6.75	0.307	16.25	0.125	25.75	0.035
7	0.340	16.5	0.124	26	0.028
7.25	0.356	16.75	0.122	26.25	0.022
7.5	0.350	17	0.121	26.5	0.017
7.75	0.323	17.25	0.119	26.75	0.013
8	0.285	17.5	0.118	27	0.010
8.25	0.249	17.75	0.116	27.25	0.006
8.5	0.221	18	0.115	27.5	0.004
8.75	0.206	18.25	0.114	27.75	0.002
9	0.201	18.5	0.114	28	0.001
9.25	0.202	18.75	0.114	28.25	0.000

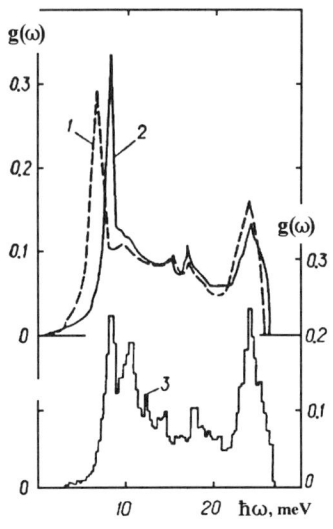
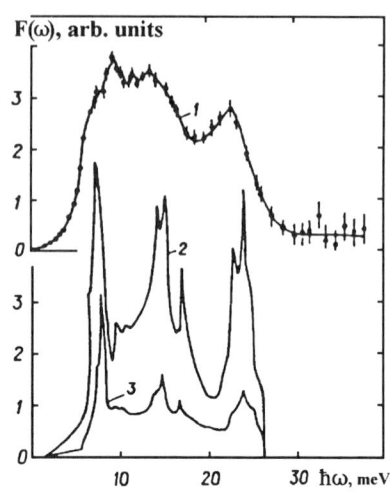

Figure 82 The EPI thermodynamic function for zinc: 1, 2 - calculations [318] for two different models, 3 - calculation [314] ($\lambda = 0.334$).

Figure 83 The phonon density of states in zinc
1 - experimental data [33], obtained by thermal neutron scattering
2, 3 - calculation [172], [284].

2.10 Cadmium

Crystal lattice: hcp
References. Point contact spectroscopy: [120, 139].
Phonon density of states: [33, 35, 55].

Figure 84
An EPI point contact spectrum for cadmium [120] (a pressed point contact of the sliding type, $R_0 = 4$ ohms, $V_{1,0} = 524$ μV, $V_{2max} = 1.19$ μV, $T = 1.5$ K).

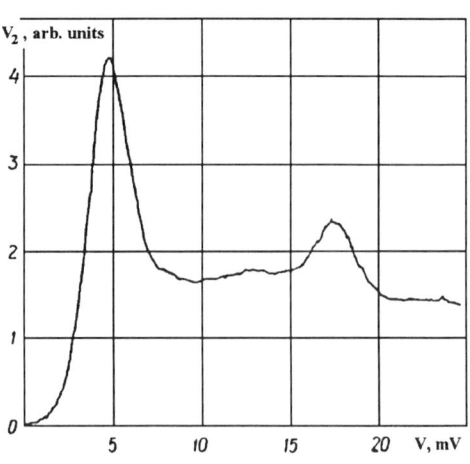

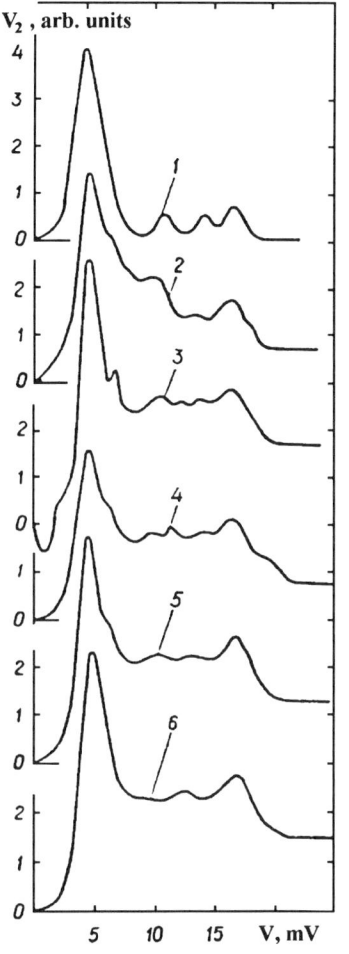
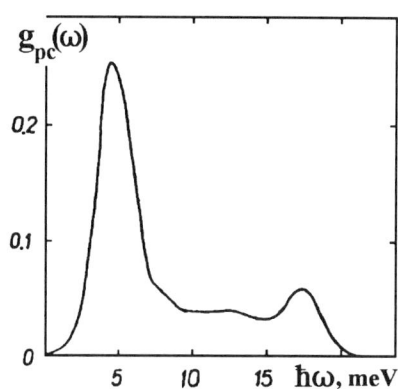
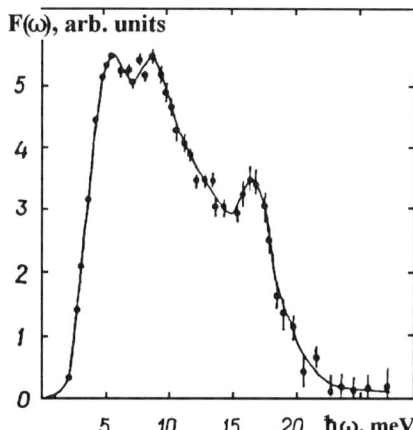

Figure 85 EPI point contact spectra for cadmium [139] (point contacts obtained by the thin film technique, T = 1.5 K, characteristics of the spectra are presented in Table 15 ; the number of a curve corresponds to the number of a contact in Table 15).

Figure 86 The EPI point contact function for cadmium, reconstructed from the point contact spectrum in Figure 84 (λ_{pc} = 0.44).

Figure 87 The phonon density of states in cadmium (experimental data [3], obtained by thermal neutron scattering).

Table 15 Characteristics of the EPI point contact spectra for cadmium shown in Figure 85.

Contact number	R_0, ohms	$V_{1,0}$, µV	Contact number	R_0, ohms	$V_{1,0}$, µV
1	200	600	4	2.18	400
2	26.1	500	5	2.7	400
3	1.88	400	6	28.3	300

Table 16 Values of the EPI point contact function for cadmium

$\hbar\omega$, meV	$g_{pc}(\omega)$	$\hbar\omega$, meV	$g_{pc}(\omega)$	$\hbar\omega$, meV	$g_{pc}(\omega)$
0	0	7.75	0.055	15.25	0.033
0.25	0.001	8	0.052	15.5	0.034
0.5	0.002	8.25	0.050	15.75	0.037
0.75	0.003	8.5	0.047	16	0.040
1	0.005	8.75	0.045	16.25	0.044
1.25	0.007	9	0.043	16.5	0.049
1.5	0.010	9.25	0.041	16.75	0.053
1.75	0.015	9.5	0.040	17	0.057
2	0.022	9.75	0.039	17.25	0.058
2.25	0.031	10	0.039	17.5	0.057
2.5	0.044	10.25	0.039	17.75	0.055
2.75	0.062	10.5	0.039	18	0.051
3	0.086	10.75	0.039	18.25	0.045
3.25	0.113	11	0.038	18.5	0.038
3.5	0.146	11.25	0.038	18.75	0.031
3.75	0.183	11.5	0.038	19	0.025
4	0.217	11.75	0.039	19.25	0.018
4.25	0.241	12	0.039	19.5	0.013
4.5	0.253	12.25	0.039	19.75	0.010
4.75	0.252	12.5	0.039	20	0.008
5	0.240	12.75	0.038	20.25	0.006
5.25	0.219	13	0.038	20.5	0.003
5.5	0.194	13.25	0.037	20.75	0.002
5.75	0.169	13.5	0.035	21	0.002
6	0.145	13.75	0.034	21.25	0.002
6.25	0.122	14	0.034	21.5	0.001
6.5	0.121	14.25	0.033	21.75	0.001
6.75	0.084	14.5	0.033	22	0.001
7	0.070	14.75	0.032	22.25	0.001
7.25	0.062	15	0.032	22.5	0.000
7.5	0.057				

2.11 Aluminum

Crystal lattice: FCC

References Point contact spectroscopy: [37, 123, 272, 292, 312].
EPI functions: [37, 165-167, 175, 180, 183, 208, 238, 241, 340].
Phonon density of states : [17, 37, 165, 175, 187, 199, 200, 207, 241, 264, 274, 285, 300, 305, 307, 327].

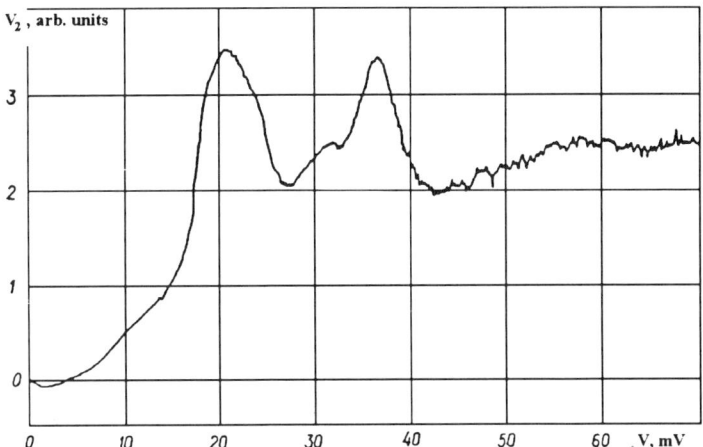

Figure 88 An EPI point contact spectrum for aluminum [122] (a pressed point contact of the sliding type, R_0 = 10 ohms, $V_{1,0}$ = 450 µV, V_{2max} = 0.57 µV, T = 1.83 K).

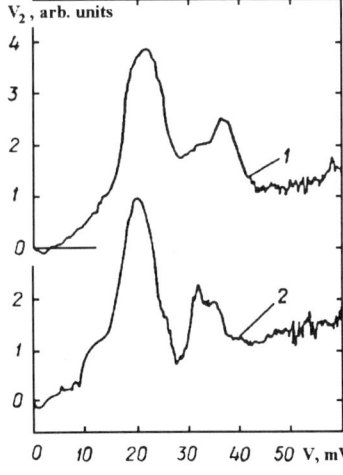

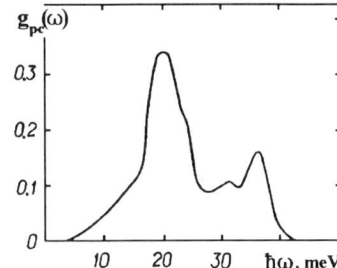

Figure 89 EPI point contact spectra for aluminum [122]:
1 - spectrum of a pressed point contact of the sliding type between films sputtered onto cylindrical substrates (R_0 = 9.6 ohms, $V_{1,0}$ = 600 V, T = 1.84 K);

2 - An atypical spectrum of a pressed point contact of the sliding type between single crystal electrodes of random orientation, reflecting possible anisotropy of the EPI (R_0 = 10 ohms, $V_{1,0}$ = 900 µV, T = 1.4 K).

Figure 90 The EPI point contact function for aluminum, reconstructed from the point contact spectrum in Figure 88 (λ_{pc} = 0.45).

Table 17 Values of the EPI point contact function for aluminum.

$\hbar\omega$, meV	$g_{pc}(\omega)$	$\hbar\omega$, meV	$g_{pc}(\omega)$	$\hbar\omega$, meV	$g_{pc}(\omega)$
0	0	15	0.113	30	0.102
0.5	0.000	15.5	0.121	30.5	0.105
1	0.000	16	0.131	31	0.106
1.5	0.001	16.5	0.150	31.5	0.105
2	0.001	17	0.195	32	0.100
2.5	0.001	17.5	0.248	32.5	0.098
3	0.001	18	0.302	33	0.100
3.5	0.002	18.5	0.317	33.5	0.109
4	0.003	19	0.329	34	0.118
4.5	0.004	19.5	0.332	34.5	0.132
5	0.006	20	0.336	35	0.144
5.5	0.008	20.5	0.333	35.5	0.157
6	0.012	21	0.322	36	0.162
6.5	0.016	21.5	0.302	36.5	0.150
7	0.021	22	0.281	37	0.138
7.5	0.026	22.5	0.260	37.5	0.113
8	0.030	23	0.238	38	0.097
8.5	0.034	23.5	0.221	38.5	0.073
9	0.040	24	0.198	39	0.051
9.5	0.045	24.5	0.171	39.5	0.037
10	0.049	25	0.139	40	0.027
10.5	0.054	25.5	0.115	40.5	0.019
11	0.060	26	0.100	41	0.011
11.5	0.066	26.5	0.091	41.5	0.007
12	0.075	27	0.086	42	0.004
12.5	0.079	27.5	0.087	42.5	0.001
13	0.084	28	0.088	43	0.000
13.5	0.091	28.5	0.092		
14	0.096	29	0.095		
14.5	0.105	29.5	0.098		

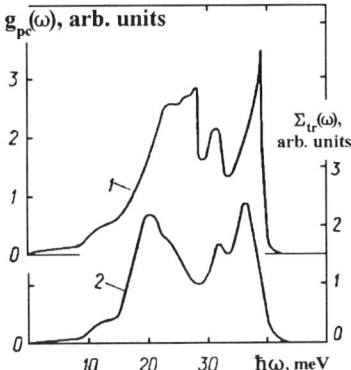
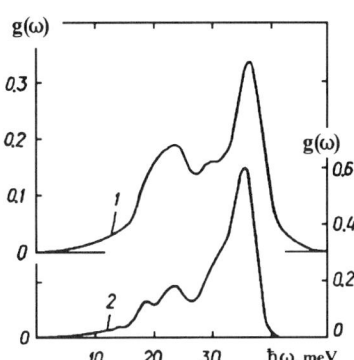

Figure 91 EPI functions for aluminum:
1 - point contact EPI function, calculated in reference [37];
2 - EPI function

$$\sum\nolimits_{tr}(\omega) = 6g_{tr\ k}^{[100]}(\omega) + 12g_{tr\ k}^{[110]}(\omega) + 8g_{tr\ k}^{[111]}(\omega),$$

calculated in reference [122] using the function $g_{tr\ k}(\omega)$ [208] (putting in broadening of 2.7 meV along with the calculation).

Figure 92 The thermodynamic EPI function for aluminum:
1 - data [180], obtained by the tunneling effect method ($\lambda = 0.381$)
2 - data [310], obtained by the proximity tunneling effect method ($\lambda = 0.42$).

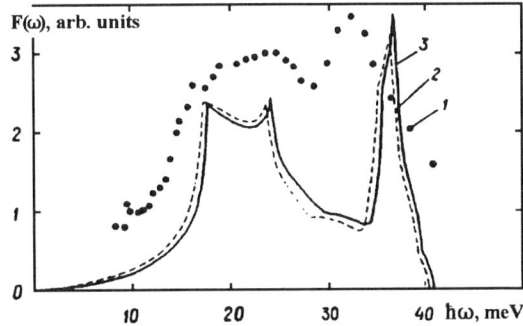

Figure 93 The phonon density of states in aluminum:
1 - experimental data [17], obtained by thermal neutron scattering;
2, 3 - calculation [199] for $T = 300$ K and $T = 80$ K.

2.12 Gallium

Crystal lattice: α-phase - rhombohedral
References. Point contact spectroscopy: [128]. Phonon density of states: [14].

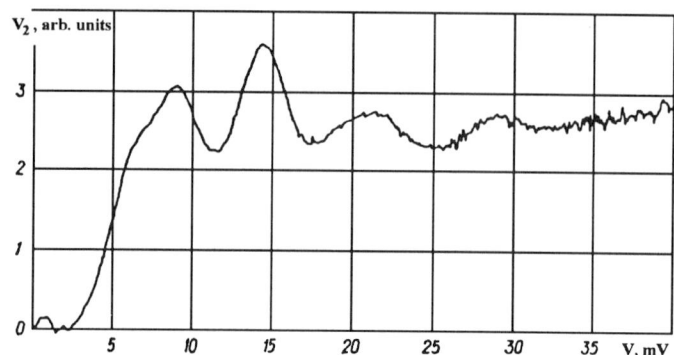

Figure 94 An EPI point contact spectrum for α-gallium [128] (a pressed point contact of the sliding type, R_0 = 6.4 ohms, $V_{1,0}$ = 618 μV, V_{2max} = 0.84 μV, T = 4.2 K, B = 3.24 T).

Figure 95 An EPI point contact spectrum in mixed α- and β-phase gallium [128] (a pressed point contact of the sliding type, R_0 = 1.5 ohms, $V_{1,0}$ = 500 μV, T = 2 K).

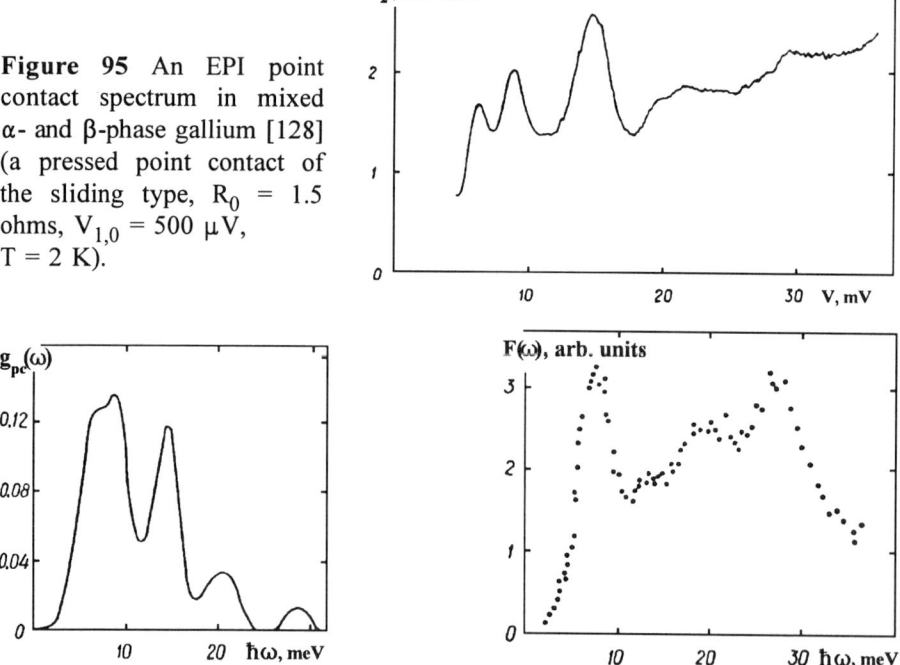

Figure 96 The EPI point contact function for α-gallium, reconstructed from the point contact spectrum in Figure 94 (λ_{pc} = 0.31).

Figure 97 The phonon density of states in α-gallium (experimental data [14], obtained by thermal neutron scattering).

Table 18 Values of the EPI point contact function for α-gallium.

$\hbar\omega$, meV	$g_{pc}(\omega)$	$\hbar\omega$, meV	$g_{pc}(\omega)$	$\hbar\omega$, meV	$g_{pc}(\omega)$
0	0	10.8	0.062	21.6	0.030
0.3	0.000	11.1	0.054	21.9	0.028
0.6	0.000	11.4	0.051	22.2	0.024
0.9	0.000	11.7	0.051	22.5	0.019
1.2	0.000	12	0.054	22.8	0.015
1.5	0.000	12.3	0.062	23.1	0.011
1.8	0.001	12.6	0.072	23.4	0.007
2.1	0.002	12.9	0.084	23.7	0.004
2.4	0.003	13.2	0.095	24	0.002
2.7	0.005	13.5	0.104	24.3	0.000
3	0.010	13.8	0.113	24.6	0.000
3.3	0.017	14.1	0.119	24.9	0.000
3.6	0.025	14.4	0.118	25.2	0.000
3.9	0.034	14.7	0.113	25.5	0.000
4.2	0.046	15	0.104	25.8	0.000
4.5	0.060	15.3	0.090	26.1	0.000
4.8	0.074	15.6	0.074	26.4	0.000
5.1	0.087	15.9	0.059	26.7	0.002
5.4	0.101	16.2	0.045	27	0.004
5.7	0.112	16.5	0.034	27.3	0.007
6	0.120	16.8	0.026	27.6	0.009
6.3	0.125	17.1	0.020	27.9	0.011
6.6	0.127	17.4	0.018	28.2	0.013
6.9	0.128	17.7	0.018	28.5	0.013
7.2	0.128	18	0.019	28.8	0.013
7.5	0.129	18.3	0.021	29.1	0.013
7.8	0.132	18.6	0.024	29.4	0.012
8.1	0.135	18.9	0.026	29.7	0.011
8.4	0.137	19.2	0.027	30	0.009
8.7	0.136	19.5	0.029	30.3	0.007
9	0.132	19.8	0.030	30.6	0.005
9.3	0.124	20.1	0.032	30.9	0.002
9.6	0.113	20.4	0.033	31.2	0.001
9.9	0.099	20.7	0.033	31.5	0.001
10.2	0.086	21	0.032	31.8	0.001
10.5	0.073	21.3	0.031	32.1	0.000

2.13 Indium

Crystal lattice: face-centered tetragonal

References. Point contact spectroscopy: [112, 137, 142, 143].
EPI functions: [39, 75, 76, 90, 182].
Phonon density of states: [193, 280, 285].

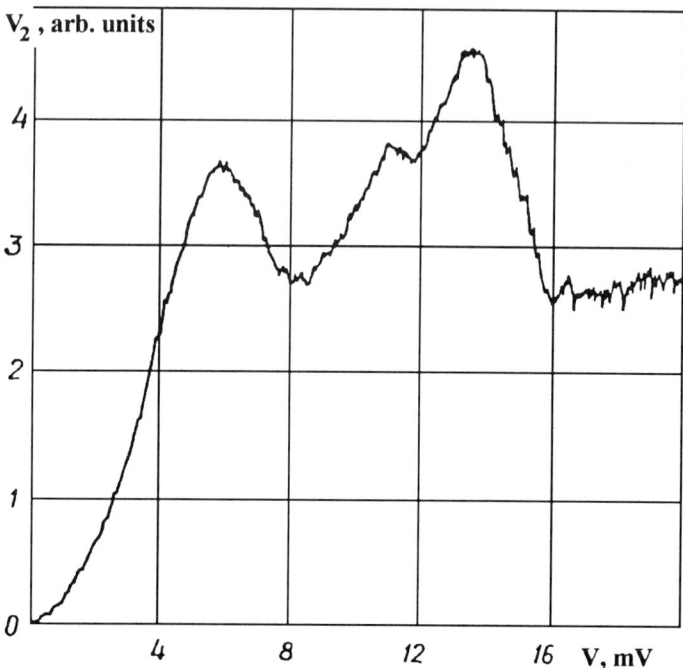

Figure 98 An EPI point contact spectrum for indium [112] (a pressed point contact of the sliding type, $R_0 = 6.3$ ohms, $V_{1,0} = 418$ μV, $V_{2max} = 1.07$ μV, $T = 1.55$ K, $B = 7.8 \times 10^{-2}$ T).

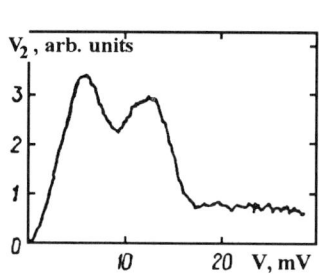 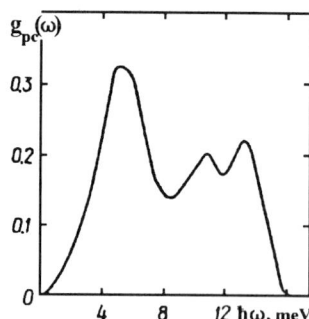

Figure 99 An EPI point contact spectrum for indium [143] (a point contact obtained by the thin film technique, $R_0 = 37.5$ ohms, $V_{1,0} = 380$ μV, $T = 4.2$ K).

Figure 100 The EPI point contact function for indium, reconstructed from the point contact spectrum in Figure 98 ($\lambda_{pc} = 0.85$).

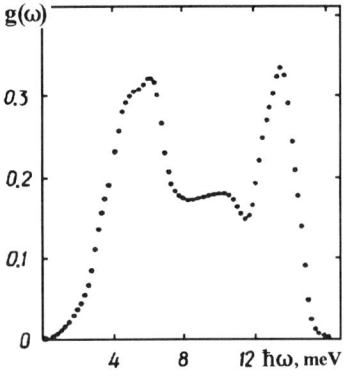

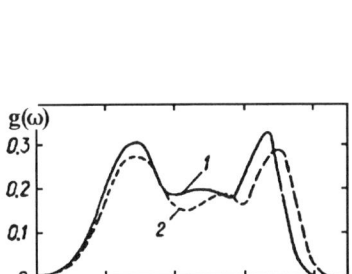

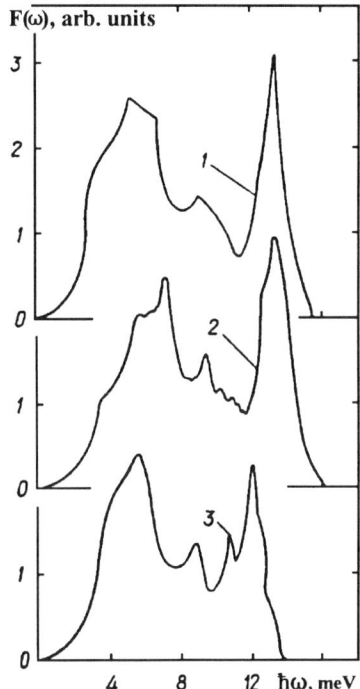

Figure 101 The EPI thermodynamic function for indium [182], obtained by the tunnel effect method ($\lambda = 0.834$).

Figure 102 Dependence of the EPI thermodynamic function for indium on hydrostatic pressure (data [90], obtained by the tunnel effect method): 1 - P = 0, $\lambda = 0.805$; 2 - P = 10 kBar.

Figure 103 The phonon density of states in indium:
1 - calculation of Smith and Richardt (cited in [228]):
2, 3 - calculations [193], [285].

Table 19 Values of the EPI point contact function for indium.

$\hbar\omega$, meV	$g_{pc}(\omega)$	$\hbar\omega$, meV	$g_{pc}(\omega)$	$\hbar\omega$, meV	$g_{pc}(\omega)$
0	0	1.5	0.040	3	0.140
0.25	0.002	1.75	0.053	3.25	0.161
0.5	0.006	2	0.067	3.5	0.186
0.75	0.012	2.25	0.081	3.75	0.214
1	0.019	2.5	0.098	4	0.238
1.25	0.029	2.75	0.120	4.25	0.257

Table 19, contd.

$\hbar\omega$, meV	$g_{pc}(\omega)$	$\hbar\omega$, meV	$g_{pc}(\omega)$	$\hbar\omega$, meV	$g_{pc}(\omega)$
4.5	0.287	8.75	0.143	12.75	0.208
4.75	0.307	9	0.149	13	0.217
5	0.326	9.25	0.155	13.25	0.222
5.25	0.325	9.5	0.162	13.5	0.216
5.5	0.324	9.75	0.170	13.75	0.198
5.75	0.317	10	0.178	14	0.174
6	0.302	10.25	0.185	14.25	0.148
6.25	0.282	10.5	0.194	14.5	0.125
6.5	0.263	10.75	0.204	14.75	0.105
6.75	0.244	11	0.202	15	0.084
7	0.213	11.25	0.191	15.25	0.058
7.25	0.182	11.5	0.180	15.5	0.028
7.5	0.165	11.75	0.174	15.75	0.008
7.75	0.155	12	0.176	16	0.002
8	0.147	12.25	0.186	16.25	0.001
8.25	0.140	12.5	0.198	16.5	0.000
8.5	0.138				

Table 20 Values of the EPI thermodynamic function for indium [90].

$\hbar\omega$, meV	$g_{pc}(\omega)$	$\hbar\omega$, meV	$g_{pc}(\omega)$	$\hbar\omega$, meV	$g_{pc}(\omega)$
0	0	5.4	0.307	10.8	0.191
0.2	0.000	5.6	0.308	11	0.186
0.4	0.001	5.8	0.309	11.2	0.182
0.6	0.003	6	0.310	11.4	0.182
0.8	0.006	6.2	0.312	11.6	0.188
1	0.009	6.4	0.307	11.8	0.200
1.2	0.012	6.6	0.289	12	0.223
1.4	0.017	6.8	0.265	12.2	0.247
1.6	0.022	7	0.240	12.4	0.266
1.8	0.028	7.2	0.219	12.6	0.284
2	0.035	7.4	0.206	12.8	0.298
2.2	0.042	7.6	0.198	13	0.311
2.4	0.051	7.8	0.193	13.2	0.327
2.6	0.065	8	0.190	13.4	0.337
2.8	0.083	8.2	0.192	13.6	0.322
3	0.104	8.4	0.195	13.8	0.292
3.2	0.125	8.6	0.198	14	0.257
3.4	0.144	8.8	0.202	14.2	0.221
3.6	0.160	9	0.203	14.4	0.183
3.8	0.176	9.2	0.204	14.6	0.143
4	0.194	9.4	0.204	14.8	0.100
4.2	0.217	9.6	0.205	15	0.064
4.4	0.242	9.8	0.205	15.2	0.041
4.6	0.265	10	0.204	15.4	0.026
4.8	0.281	10.2	0.203	15.6	0.006
5	0.294	10.4	0.200	15.8	0.000
5.2	0.303	10.6	0.196		

2.14 Thallium

Crystal lattice: α-phase - hcp

References. Point contact spectroscopy: [70]. EPI functions: [182, 317].
Phonon density of states: [317, 334].

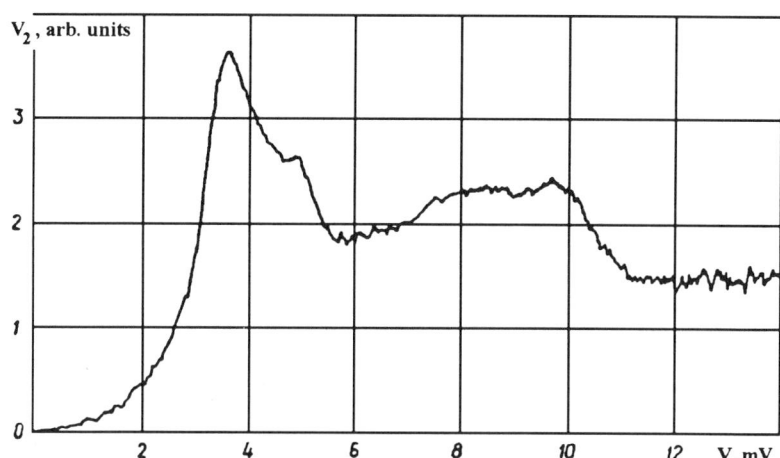

Figure 104 An EPI point contact spectrum for α-thallium [70] (a pressed point contact of the sliding type, R_0 = 4.2 ohms, $V_{1,0}$ = 357 μV, V_{2max} = 1.02 μV, T = 1.5 K, B = 0.122 T).

Table 21 Values of the EPI point contact function for thallium.

ℏω, meV	$g_{pc}(\omega)$	ℏω, meV	$g_{pc}(\omega)$	ℏω, meV	$g_{pc}(\omega)$
0	0	4	0.407	8	0.164
0.2	0.002	4.2	0.356	8.2	0.164
0.4	0.003	4.4	0.317	8.4	0.163
0.6	0.004	4.6	0.288	8.6	0.157
0.8	0.009	4.8	0.280	8.8	0.152
1	0.010	5	0.268	9	0.143
1.2	0.015	5.2	0.214	9.2	0.140
1.4	0.024	5.4	0.165	9.4	0.144
1.6	0.033	5.6	0.134	9.6	0.151
1.8	0.047	5.8	0.129	9.8	0.148
2	0.061	6	0.129	10	0.134
2.2	0.085	6.2	0.132	10.2	0.114
2.4	0.112	6.4	0.132	10.4	0.084
2.6	0.151	6.6	0.133	10.6	0.055
2.8	0.190	6.8	0.135	10.8	0.035
3	0.263	7	0.141	11	0.018
3.2	0.389	7.2	0.146	11.2	0.005
3.4	0.492	7.4	0.157	11.4	0.003
3.6	0.514	7.6	0.163	11.6	0.001
3.8	0.462	7.8	0.165	11.8	0.000

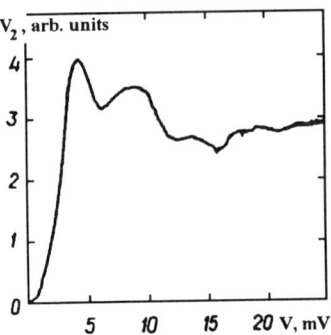
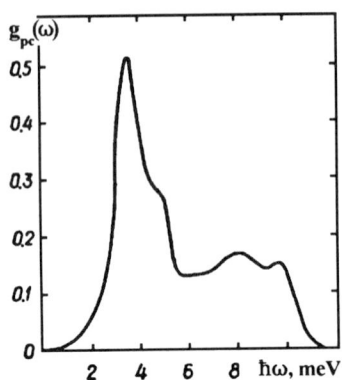

Figure 105 An EPI point contact spectrum for α-thallium [70] (a pressed point contact of the sliding type, $R_0 = 8.35$ ohms, $V_{1,0} = 650$ μV, $T = 1.5$ K, $B = 0.109$ T).

Figure 106 The EPI point contact function for α-thallium, reconstructed from the point contact spectrum in Figure 104 ($\lambda_{pc} = 0.78$).

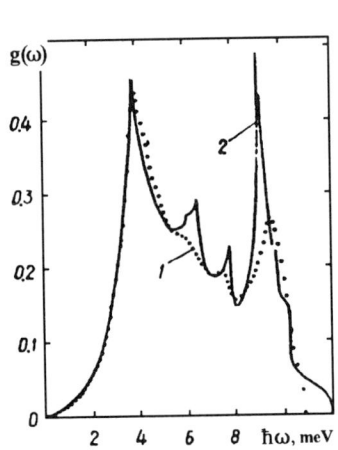

Figure 107 The EPI thermodynamic function for α-thallium:
 1 - data [182], obtained by the tunnel effect method ($\lambda = 0.780$);
 2 - calculation [317].

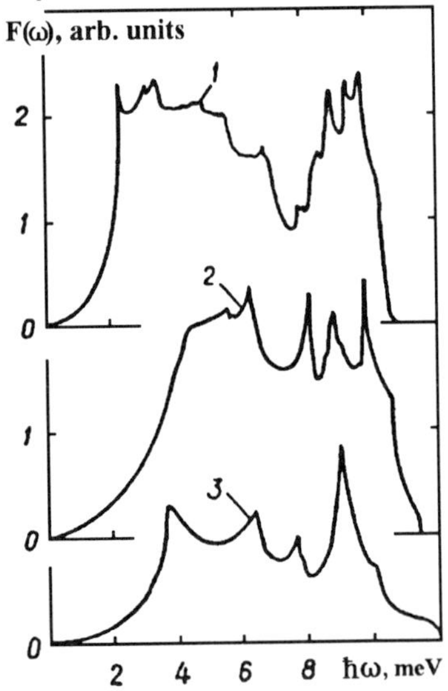

Figure 108 The phonon density of states in thallium:
 1, 2 - calculations [334] for $T = 286$ K and $T = 77$ K;
 3 - calculation [317].

2.15 Tin

Crystal lattice: β-phase - body-centered tetragonal
References. Point contact spectroscopy: [117, 119, 136, 142, 145].
EPI functions: [39, 40, 232, 233, 290].
Phonon density of states: [18, 56, 226, 227, 258].

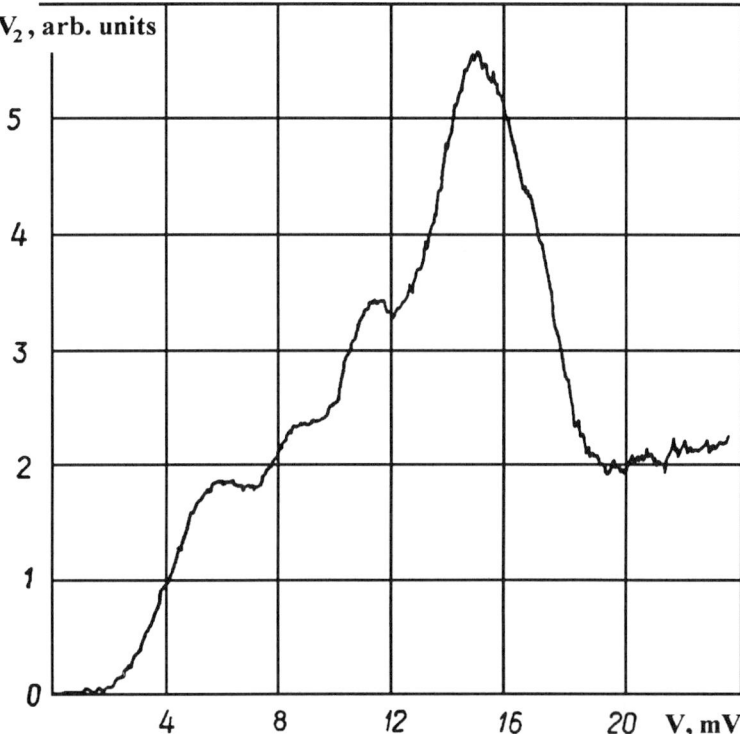

Figure 109 An EPI point contact spectrum for β-tin [117] (a pressed point contact of the sliding type, R_0 = 5.7 ohms, $V_{1,0}$ = 431 μV, V_{2max} = 1.22 μV, T = 1.6 K, B = 0.094 T).

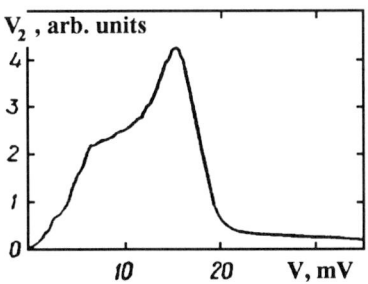

Figure 110
An EPI point contact spectrum for β-tin [142] (a point contact obtained by the thin film technique, R_0 = 85 ohms, $V_{1,0}$ = 600 μV, T = 4.2 K).

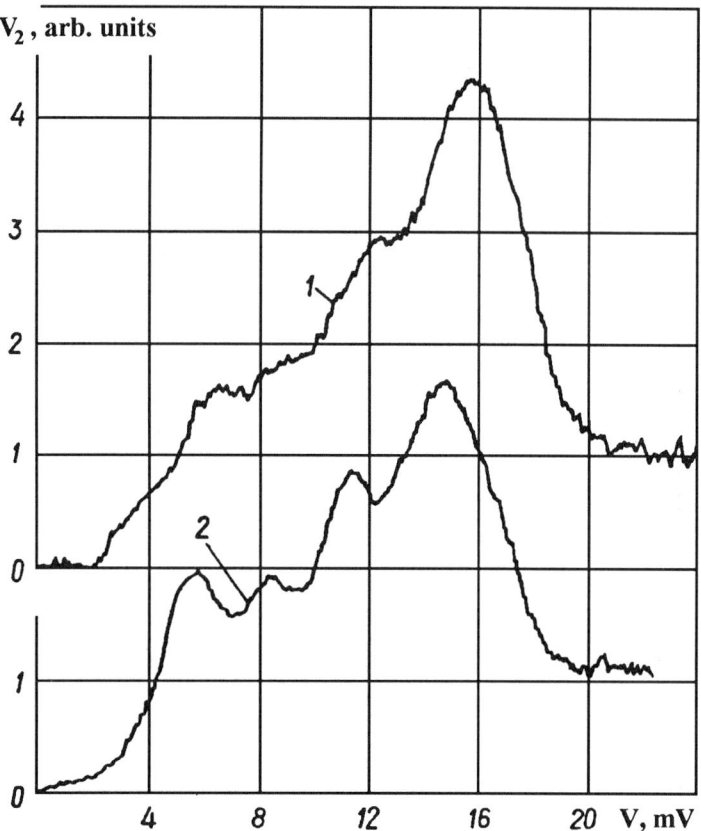

Figure 111 Anisotropy of the EPI point contact spectra for β-tin [117] in the [100] (1) and [001] (2) directions (a pressed point contact of the sliding type):
1 - $R_0 = 17.8$ ohms, $V_{1,0} = 789$ μV, $V_{2max} = 1.03$ μV, $T = 1.7$ K, $B = 0.151$ T;
2 - $R_0 = 5.8$ ohms, $V_{1,0} = 549$ μV, $V_{2max} = 0.88$ μV, $T = 1.6$ K, $B = 0.094$ T.

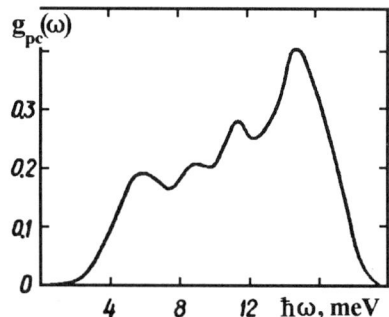

Figure 112 The EPI point contact function in β-tin, reconstructed from the point contact spectrum in Figure 109 ($\lambda_{pc} = 0.71$).

Table 22 Values of the EPI point contact function for β-tin.

ℏω, meV	$g_{pc}(\omega)$	ℏω, meV	$g_{pc}(\omega)$	ℏω, meV	$g_{pc}(\omega)$
0	0	6.75	0.174	13.25	0.283
0.25	0.000	7	0.169	13.5	0.301
0.5	0.000	7.25	0.166	13.75	0.325
0.75	0.001	7.5	0.168	14	0.352
1	0.001	7.75	0.174	14.25	0.378
1.25	0.002	8	0.184	14.5	0.397
1.5	0.003	8.25	0.194	14.75	0.406
1.75	0.004	8.5	0.202	15	0.401
2	0.007	8.75	0.205	15.25	0.386
2.25	0.011	9	0.205	15.5	0.365
2.5	0.017	9.25	0.204	15.75	0.345
2.75	0.026	9.5	0.202	16	0.323
3	0.037	9.75	0.201	16.25	0.297
3.25	0.050	10	0.206	16.5	0.265
3.5	0.065	10.25	0.221	16.75	0.237
3.75	0.083	10.5	0.241	17	0.210
4	0.102	10.75	0.258	17.25	0.181
4.25	0.119	11	0.271	17.5	0.149
4.5	0.136	11.25	0.280	17.75	0.115
4.75	0.155	11.5	0.279	18	0.082
5	0.172	11.75	0.267	18.25	0.053
5.25	0.183	12	0.254	18.5	0.031
5.5	0.188	12.25	0.251	18.75	0.018
5.75	0.190	12.5	0.255	19	0.010
6	0.188	12.75	0.261	19.25	0.005
6.25	0.185	13	0.270	19.5	0.000
6.5	0.180				

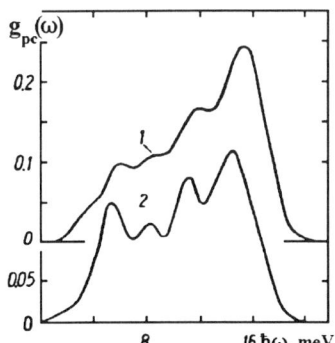
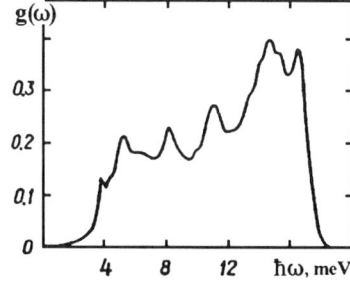

Figure 113 Anisotropy of the EPI point contact function for β-tin (reconstructed from the corresponding point contact spectra in Figure 111):
1 - [100] direction , 2 - [001].

Figure 114 The EPI thermodynamic function for β-tin [290], obtained by the tunnel effect method ($\lambda = 0.72$).

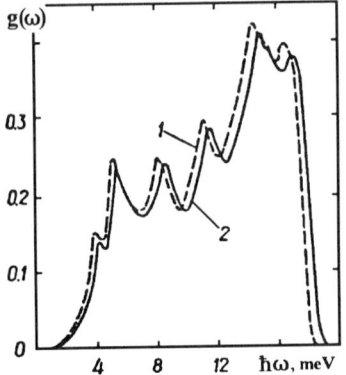

Figure 115 Dependence of the EPI thermodynamic function for β-tin on hydrostatic pressure (data [39], obtained by the tunnel effect method): 1 - P = 0 ; 2 - P = 10 kBar.

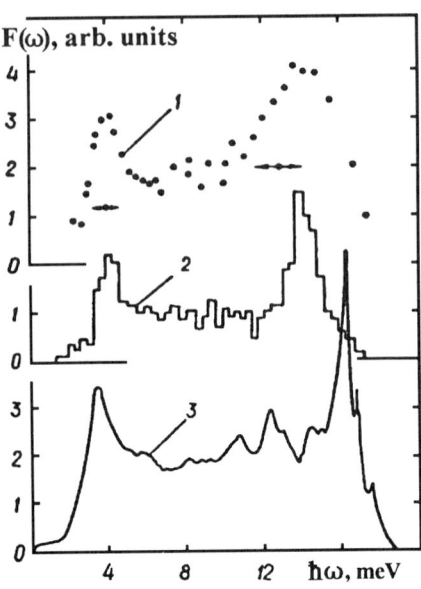

Figure 116 The phonon density of states in β-tin: 1 - data [56], obtained by thermal neutron scattering (the pointers show the resolution of the experiment); 2, 3 - calculations [18],[227].

2.16 Lead

Crystal lattice: FCC
References. Point contact spectroscopy: [116, 118, 141, 142].
EPI functions: [21, 39, 40, 95-97, 146, 163, 175, 184, 189, 232, 233, 240, 252, 289, 311]. **Phonon density of states:** [57, 175, 176, 198, 203, 291, 307].

Table 23 Values of the EPI point contact function for lead.

$\hbar\omega$, meV	$g_{pc}(\omega)$	$\hbar\omega$, meV	$g_{pc}(\omega)$	$\hbar\omega$, meV	$g_{pc}(\omega)$
0	0	2.4	0.111	4.8	0.815
0.2	0.001	2.6	0.143	5	0.756
0.4	0.002	2.8	0.203	5.2	0.700
0.6	0.004	3	0.297	5.4	0.645
0.8	0.006	3.2	0.375	5.6	0.585
1	0.009	3.4	0.502	5.8	0.529
1.2	0.018	3.6	0.631	6	0.482
1.4	0.029	3.8	0.711	6.2	0.425
1.6	0.041	4	0.767	6.4	0.376
1.8	0.052	4.2	0.820	6.6	0.356
2	0.064	4.4	0.847	6.8	0.363
2.2	0.083	4.6	0.846	7	0.382

Table 23, contd.

ℏω, meV	$g_{pc}(\omega)$	ℏω, meV	$g_{pc}(\omega)$	ℏω, meV	$g_{pc}(\omega)$
7.2	0.412	9	0.868	10.6	0.096
7.4	0.461	9.2	0.781	10.8	0.062
7.6	0.538	9.4	0.666	11	0.040
7.8	0.631	9.6	0.531	11.2	0.031
8	0.733	9.8	0.417	11.4	0.023
8.2	0.854	10	0.327	11.6	0.012
8.4	0.912	10.2	0.230	11.8	0.002
8.6	0.961	10.4	0.150	12	0.000
8.8	0.971				

Table 24 Values of the EPI thermodynamic function for lead [311] ($\lambda = 1.538$).

ℏω, meV	$g_{pc}(\omega)$	ℏω, meV	$g_{pc}(\omega)$	ℏω, meV	$g_{pc}(\omega)$
0	0	3.5	0.6459	7	0.3183
0.1	0.0002	3.6	0.7304	7.1	0.3253
0.2	0.0007	3.7	0.7946	7.2	0.3393
0.3	0.0016	3.8	0.8330	7.3	0.3581
0.4	0.0029	3.9	0.8585	7.4	0.3829
0.5	0.0045	4	0.8863	7.5	0.4251
0.6	0.0064	4.1	0.9223	7.6	0.4916
0.7	0.0087	4.2	0.9619	7.7	0.5683
0.8	0.0114	4.3	0.9854	7.8	0.6450
0.9	0.0145	4.4	0.9788	7.9	0.7357
1	0.0178	4.5	0.9526	8	0.8529
1.1	0.0216	4.6	0.9195	8.1	0.9893
1.2	0.0257	4.7	0.8803	8.2	1.1174
1.3	0.0301	4.8	0.8337	8.3	1.1850
1.4	0.0349	4.9	0.7788	8.4	1.1541
1.5	0.0401	5	0.7177	8.5	1.0264
1.6	0.0456	5.1	0.6592	8.6	0.8031
1.7	0.0515	5.2	0.6099	8.7	0.6188
1.8	0.0577	5.3	0.5652	8.8	0.4364
1.9	0.0643	5.4	0.5199	8.9	0.2892
2	0.0713	5.5	0.4762	9	0.1758
2.1	0.0786	5.6	0.4369	9.1	0.0979
2.2	0.0858	5.7	0.4039	9.2	0.0537
2.3	0.0951	5.8	0.3779	9.3	0.0338
2.4	0.1074	5.9	0.3576	9.4	0.0253
2.5	0.1200	6	0.3408	9.5	0.0216
2.6	0.1332	6.1	0.3267	9.6	0.0187
2.7	0.1469	6.2	0.3164	9.7	0.0176
2.8	0.1654	6.3	0.3146	9.8	0.0216
2.9	0.1960	6.4	0.3242	9.9	0.0252
3	0.2441	6.5	0.3380	10	0.0204
3.1	0.3071	6.6	0.3468	10.1	0.0129
3.2	0.3808	6.7	0.3453	10.2	0.0137
3.3	0.4639	6.8	0.3336	10.3	0.0000
3.4	0.5541	6.9	0.3212		

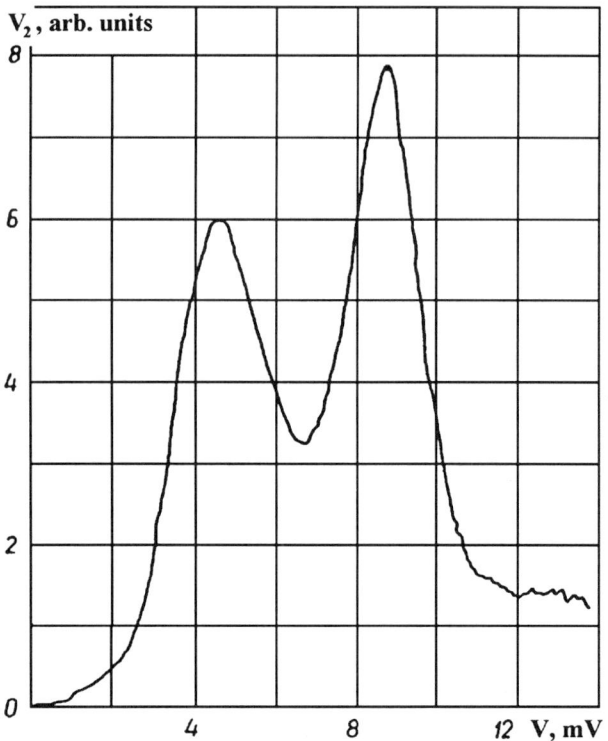

Figure 117 An EPI point contact spectrum for lead [142] (a point contact obtained by the thin film technique, $R_0 = 83$ ohms, $V_{1,0} = 250$ μV, $V_{2max} = 0.2$ μV, $T = 1.9$ K, $B_\perp = 1$ T).

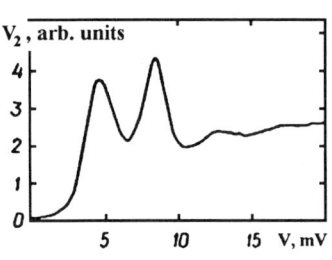
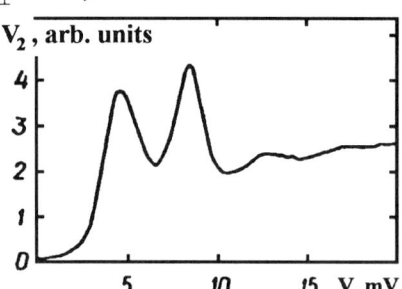

Figure 118 An EPI point contact spectrum for lead [142] (a point contact obtained by the thin film technique, $R_0 = 1.88$ ohms, $V_{1,0} = 395$ μV, $T = 2.5$ K, $B_\perp = 0.413$ T).

Figure 119 The EPI point contact function for lead, reconstructed from the point contact spectrum in Figure 117 ($\lambda_{pc} = 1.7$).

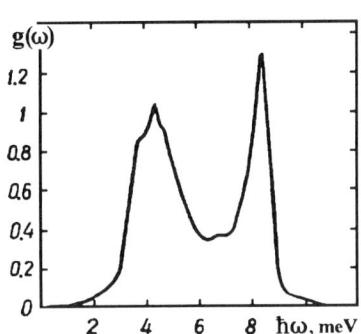
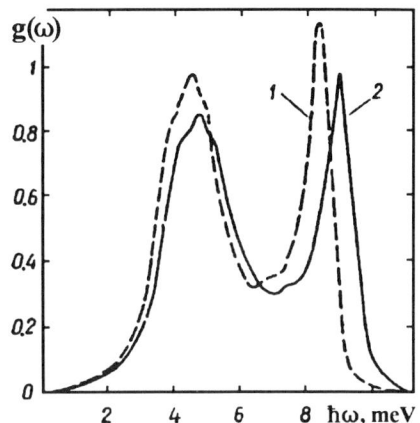

Figure 120 The EPI thermodynamic function for lead [289], obtained by the tunnel effect method ($\lambda = 1.55$).

Figure 121 Dependence of the EPI thermodynamic function for lead on hydrostatic pressure (data [95], obtained by the tunnel effect method):
1 - P = 0 ; 2 - P = 12.2 kBar.

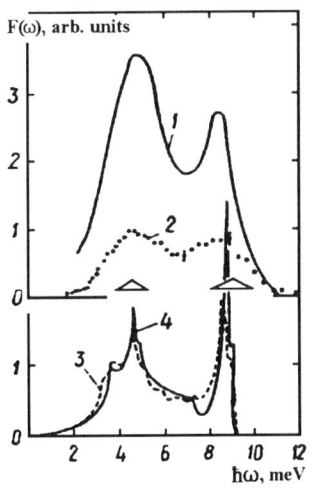

Figure 122 The phonon density of states in lead: 1, 2 - data, [291] and [57], obtained by thermal neutron scattering (the triangles show the resolution of the experimental setup [57]); 3, 4 - calculations [307], [176].

2.17 Vanadium

Crystal lattice: BCC
References. Point contact spectroscopy: [92].
EPI functions: [340]. Phonon density of states: [45, 101, 121, 171, 173, 174, 185, 212, 228, 265, 268, 278, 297, 300, 303].

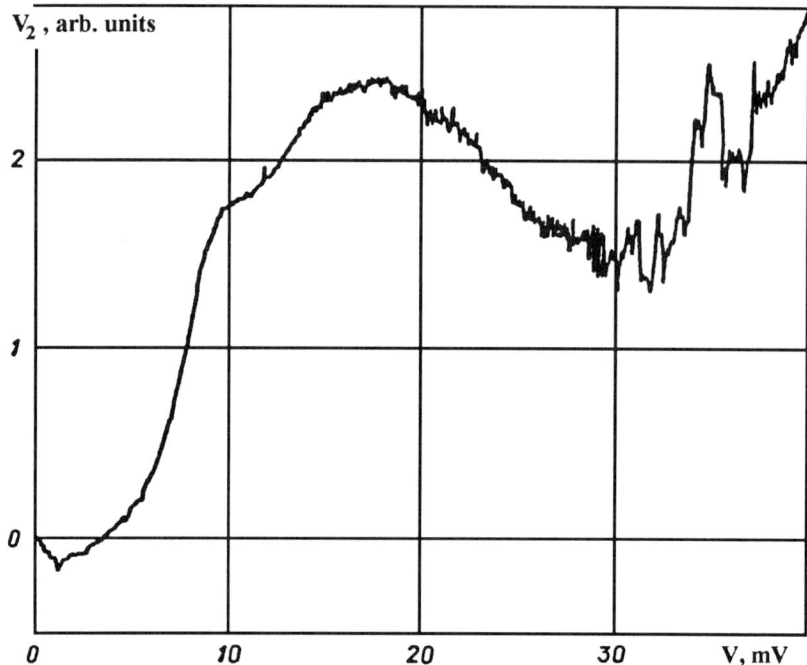

Figure 123 An EPI point contact spectrum for vanadium [92] (a pressed needle-plane point contact, $R_0 = 17$ ohms, $V_{1,0} = 2200$ μV, $V_{2max} = 0.62$ μV, T = 4.2 K, B = 1.3 T).

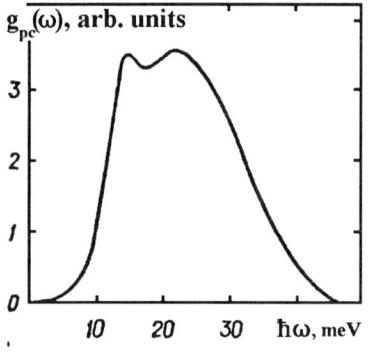 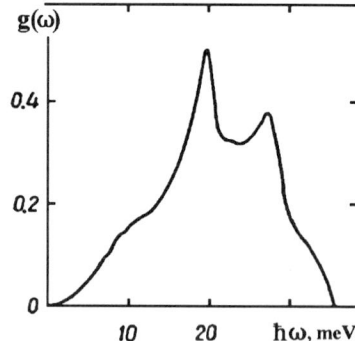

Figure 124 The EPI point contact function for vanadium, reconstructed from the point contact spectrum in Figure 123.

Figure 125 The EPI thermodynamic function for vanadium [340], obtained by the proximity tunnel effect method ($\lambda = 0.83$).

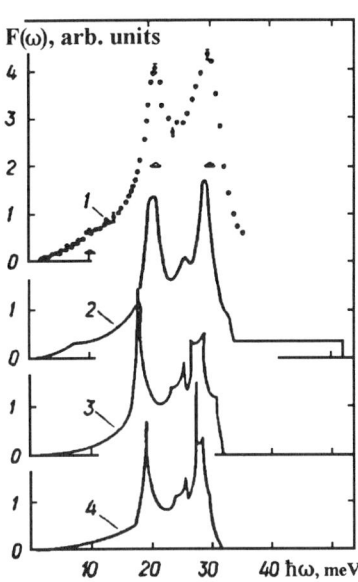

Figure 126 The phonon density of states in vanadium: 1, 2 - data [101] and [265], obtained by thermal neutron scattering (the triangles show the resolution of the experimental arrangement of [101]); 3, 4 - calculations [174] for two different models.

Table 25 Values of the EPI point contact function for vanadium.

$\hbar\omega$, meV	$\dfrac{g_{pc}(\omega)}{[g_{pc}(\omega)]_{max}}$	$\hbar\omega$, meV	$\dfrac{g_{pc}(\omega)}{[g_{pc}(\omega)]_{max}}$	$\hbar\omega$, meV	$\dfrac{g_{pc}(\omega)}{[g_{pc}(\omega)]_{max}}$
0	0	11	0.940	21.5	0.564
0.5	0.000	11.5	0.929	22	0.524
1	0.000	12	0.933	22.5	0.472
1.5	0.000	12.5	0.945	23	0.426
2	0.001	13	0.962	23.5	0.380
2.5	0.005	13.5	0.980	24	0.344
3	0.011	14	0.994	24.5	0.295
3.5	0.020	14.5	1	25	0.261
4	0.034	15	0.998	25.5	0.223
4.5	0.054	15.5	0.987	26	0.191
5	0.082	16	0.969	26.5	0.164
5.5	0.123	16.5	0.946	27	0.138
6	0.183	17	0.919	27.5	0.113
6.5	0.270	17.5	0.891	28	0.090
7	0.382	18	0.862	28.5	0.068
7.5	0.511	18.5	0.829	29	0.050
8	0.652	19	0.792	29.5	0.033
8.5	0.790	19.5	0.749	30	0.021
9	0.904	20	0.703	30.5	0.010
9.5	0.970	20.5	0.655	31	0.003
10	0.983	21	0.608	31.5	0.000
10.5	0.963				

2.18 Niobium

Crystal lattice: BCC
References. Point contact spectroscopy: [6, 134, 135, 168, 336].
 EPI functions: [89, 147, 161, 194, 286, 294, 330-332].
 Phonon density of states: [54, 259, 277, 287, 301].

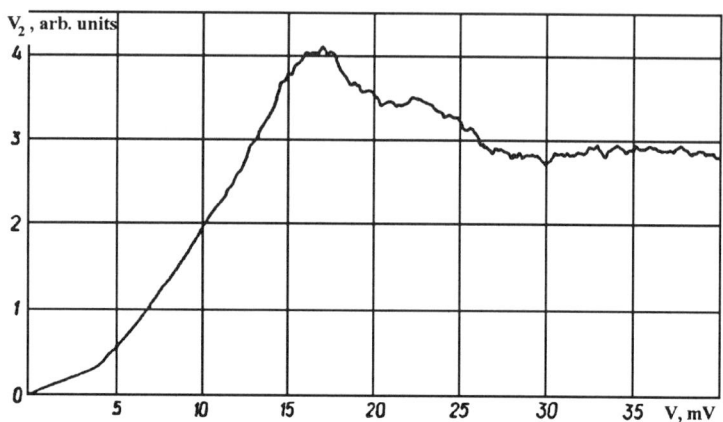

Figure 127 An EPI point contact spectrum for niobium [135] (a pressed point contact of the sliding type, R_0 = 100 ohms, $V_{1,0}$ = 950 μV, V_{2max} = 2.76 μV, T = 4.2 K, B = 5.44 T).

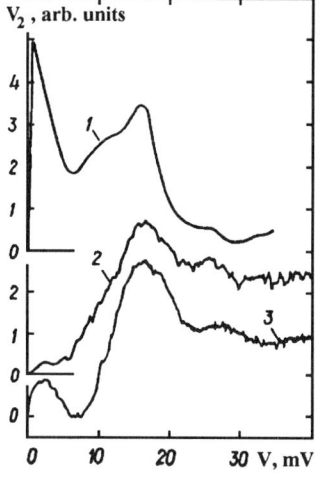

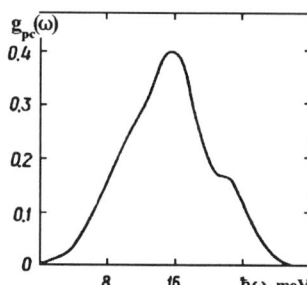

Figure 128 EPI point contact spectra for niobium: 1 - data [168], obtained by averaging over 7 spectra for a pressed needle-plane point contact ($V_{1,0}$ = 900 μV, T = 1.5 K, B = 2 T); 2, 3 - data [135] (pressed point contacts of the sliding type, T = 4.2 K; 2 - R_0 = 75 ohms, $V_{1,0}$ = 1377 μV, B = 5.6 T; 3 - R_0 = 185 ohms, $V_{1,0}$ = 2460 μV, B = 4.8 T).

Figure 129 The EPI point contact function for niobium, reconstructed from the point contact spectrum in Figure 127 (λ_{pc} = 0.81).

Table 26 Values of the EPI point contact function for niobium.

$\hbar\omega$, meV	$g_{pc}(\omega)$	$\hbar\omega$, meV	$g_{pc}(\omega)$	$\hbar\omega$, meV	$g_{pc}(\omega)$
0	0	10.5	0.254	20.5	0.186
0.5	0.005	11	0.267	21	0.174
1	0.009	11.5	0.285	21.5	0.172
1.5	0.013	12	0.296	22	0.170
2	0.019	12.5	0.310	22.5	0.167
2.5	0.021	13	0.334	23	0.154
3	0.023	13.5	0.354	23.5	0.138
3.5	0.031	14	0.370	24	0.120
4	0.038	14.5	0.394	24.5	0.102
4.5	0.052	15	0.406	25	0.089
5	0.062	15.5	0.407	25.5	0.070
5.5	0.082	16	0.405	26	0.054
6	0.098	16.5	0.397	26.5	0.037
6.5	0.116	17	0.376	27	0.028
7	0.131	17.5	0.353	27.5	0.019
7.5	0.148	18	0.315	28	0.014
8	0.164	18.5	0.283	28.5	0.008
8.5	0.184	19	0.249	29	0.003
9	0.203	19.5	0.226	29.5	0.002
9.5	0.221	20	0.204	30	0.000
10	0.241				

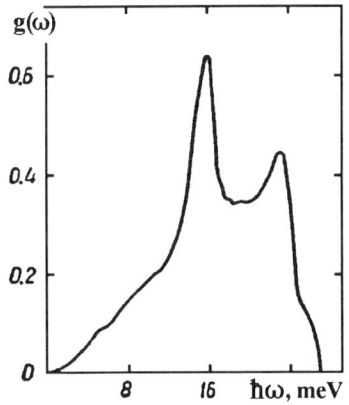

Figure 130 The EPI thermodynamic function for niobium [332], obtained by the proximity tunnel effect method ($\lambda = 1.04$).

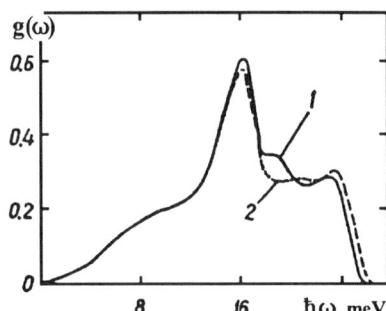

Figure 131 Dependence of the EPI thermodynamic function for niobium on hydrostatic pressure (data [89], obtained by the tunnel effect method): 1 - P = 0, $\lambda = 0.85$; 2 - P = 6 kBar.

Figure 132 The phonon density of states in niobium (calculation [277]).

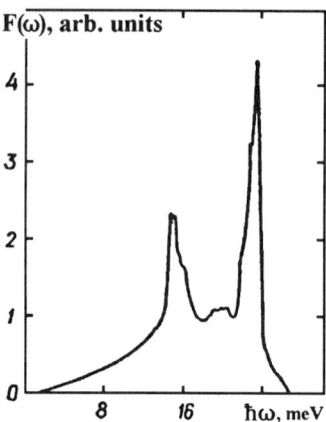

Table 27 Values of the EPI thermodynamic function for niobium [89].

ℏω, meV	g(ω)	ℏω, meV	g(ω)	ℏω, meV	g(ω)
0	0	8.7	0.169	17.4	0.353
0.3	0.000	9	0.171	17.7	0.337
0.6	0.001	9.3	0.179	18	0.339
0.9	0.003	9.6	0.175	18.3	0.341
1.2	0.005	9.9	0.184	18.6	0.332
1.5	0.008	10.2	0.189	18.9	0.326
1.8	0.012	10.5	0.196	19.2	0.315
2.1	0.016	10.8	0.210	19.5	0.289
2.4	0.021	11.1	0.206	19.8	0.282
2.7	0.027	11.4	0.209	20.1	0.280
3	0.033	11.7	0.215	20.4	0.274
3.3	0.040	12	0.227	20.7	0.267
3.6	0.047	12.3	0.234	21	0.268
3.9	0.053	12.6	0.247	21.3	0.268
4.2	0.063	12.9	0.274	21.6	0.266
4.5	0.073	13.2	0.301	21.9	0.264
4.8	0.073	13.5	0.324	22.2	0.262
5.1	0.083	13.8	0.353	22.5	0.258
5.4	0.093	14.1	0.390	22.8	0.273
5.7	0.096	14.4	0.425	23.1	0.288
6	0.106	14.7	0.456	23.4	0.265
6.3	0.117	15	0.496	23.7	0.243
6.6	0.122	15.3	0.530	24	0.215
6.9	0.125	15.6	0.559	24.3	0.167
7.2	0.142	15.9	0.588	24.6	0.119
7.5	0.147	16.2	0.589	24.9	0.079
7.8	0.154	16.5	0.545	25.2	0.041
8.1	0.170	16.8	0.474	25.5	0.003
8.4	0.169	17.1	0.400	25.8	0.000

2.19 Tantalum

Crystal lattice: BCC
References. Point contact spectroscopy: [91]. EPI functions: [89, 302, 329].
  Phonon density of states: [54, 194, 270, 333].

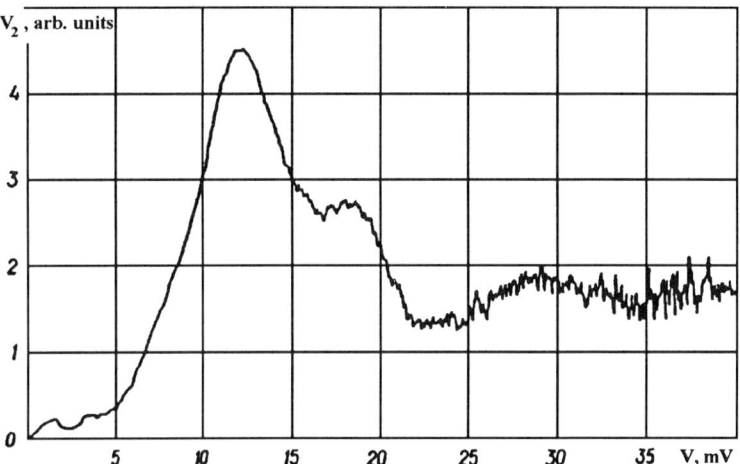

Figure 133 An EPI point contact spectrum for tantalum [92] (a pressed point contact of the sliding type, R_0 = 10 ohms, $V_{1,0}$ = 970 μV, V_{2max} = 1.69 μV, T = 4.2 K, B = 2 T).

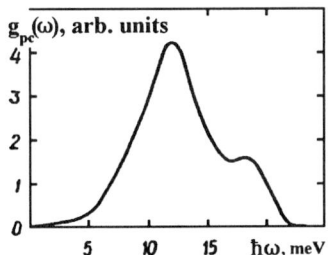
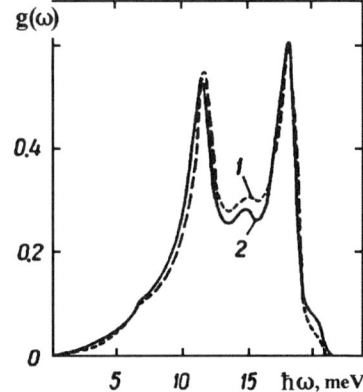

Figure 134 The EPI point contact function for tantalum, reconstructed from the point contact spectrum in Figure 133.

Figure 135 The EPI thermodynamic function for tantalum:
1 -data [302] obtained by the tunnel effect method (λ = 0.69);
2 -data [329] obtained by the proximity tunnel effect method (λ = 0.70).

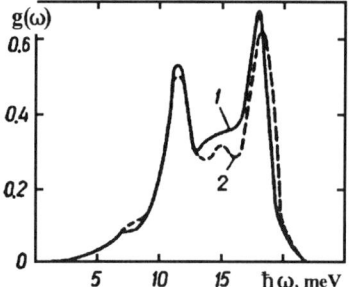

Figure 136 Dependence of the EPI thermodynamic function on hydrostatic pressure (data [89], obtained by the tunnel effect method):
1 - P = 0, λ = 0.68; 2 - P = 9 kBar.

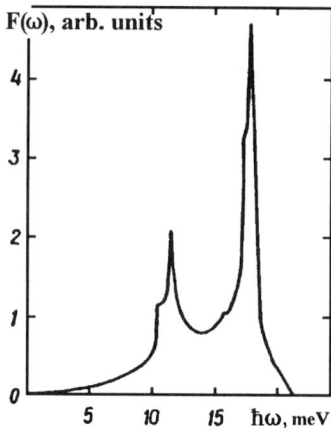

Figure 137 The phonon density of states in tantalum (calculation [33]).

Table 28 Values of the EPI point contact function for tantalum.

$\hbar\omega$, meV	$\dfrac{g_{pc}(\omega)}{[g_{pc}(\omega)]_{max}}$	$\hbar\omega$, meV	$\dfrac{g_{pc}(\omega)}{[g_{pc}(\omega)]_{max}}$	$\hbar\omega$, meV	$\dfrac{g_{pc}(\omega)}{[g_{pc}(\omega)]_{max}}$
0	0	7.8	0.388	15.3	0.475
0.3	0.004	8.1	0.427	15.6	0.446
0.6	0.007	8.4	0.461	15.9	0.417
0.9	0.011	8.7	0.495	16.2	0.392
1.2	0.014	9	0.534	16.5	0.374
1.5	0.017	9.3	0.578	16.8	0.363
1.8	0.021	9.6	0.628	17.1	0.359
2.1	0.024	9.9	0.685	17.4	0.363
2.4	0.028	10.2	0.746	17.7	0.369
2.7	0.031	10.5	0.809	18	0.372
3	0.033	10.8	0.875	18.3	0.369
3.3	0.037	11.1	0.929	18.6	0.362
3.6	0.043	11.4	0.964	18.9	0.350
3.9	0.046	11.7	0.990	19.2	0.330
4.2	0.050	12	1	19.5	0.301
4.5	0.055	12.3	0.982	19.8	0.260
4.8	0.063	12.6	0.951	20.1	0.216
5.1	0.078	12.9	0.911	20.4	0.176
5.4	0.099	13.2	0.850	20.7	0.139
5.7	0.122	13.5	0.781	21	0.102
6	0.149	13.8	0.726	21.3	0.060
6.3	0.182	14.1	0.673	21.6	0.025
6.6	0.223	14.4	0.610	21.9	0.008
6.9	0.268	14.7	0.522	22.2	0.001
7.2	0.309	15	0.507	22.5	0.000
7.5	0.347				

Table 29 Values of the EPI thermodynamic function for tantalum [89].

ℏω, meV	g(ω)	ℏω, meV	g(ω)	ℏω, meV	g(ω)
0	0	7.8	0.093	15.6	0.341
0.3	0.000	8.1	0.097	15.9	0.348
0.6	0.000	8.4	0.113	16.2	0.369
0.9	0.000	8.7	0.124	16.5	0.393
1.2	0.001	9	0.147	16.8	0.417
1.5	0.001	9.3	0.162	17.1	0.504
1.8	0.002	9.6	0.184	17.4	0.591
2.1	0.003	9.9	0.229	17.7	0.668
2.4	0.004	10.2	0.279	18	0.664
2.7	0.005	10.5	0.247	18.3	0.587
3	0.006	10.8	0.431	18.6	0.436
3.3	0.007	11.1	0.507	18.9	0.251
3.6	0.008	11.4	0.529	19.2	0.158
3.9	0.013	11.7	0.490	19.5	0.120
4.2	0.014	12	0.402	19.8	0.105
4.5	0.017	12.3	0.334	20.1	0.093
4.8	0.024	12.6	0.307	20.4	0.086
5.1	0.033	12.9	0.297	20.7	0.077
5.4	0.049	13.2	0.327	21	0.062
5.7	0.052	13.5	0.322	21.3	0.050
6	0.060	13.8	0.327	21.6	0.038
6.3	0.069	14.1	0.328	21.9	0.026
6.6	0.076	14.4	0.335	22.2	0.015
6.9	0.080	14.7	0.340	22.5	0.003
7.2	0.086	15	0.344	22.8	0.000
7.5	0.090	15.3	0.349		

2.20 Molybdenum

Crystal lattice: BCC
References. Point contact spectroscopy: [91, 141, 168, 169].
EPI functions: [276]. Phonon density of states: [55, 153, 249, 277].

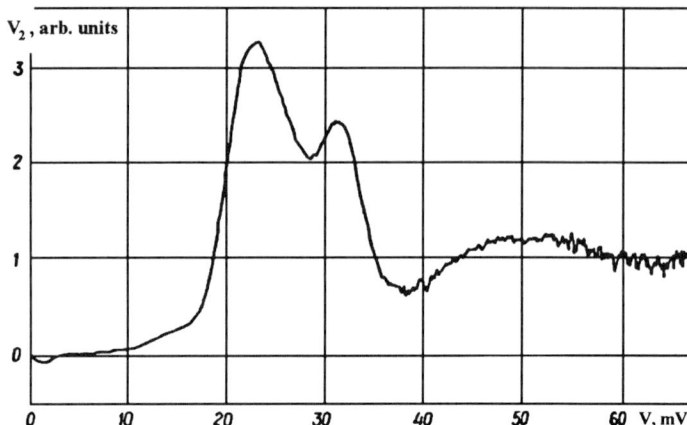

Figure 138 An EPI point contact spectrum for molybdenum [92] (a pressed point contact of the sliding type, $R_0 = 26$ ohms, $V_{1,0} = 814$ μV, $V_{2max} = 3.24$ μV, $T = 4.2$ K).

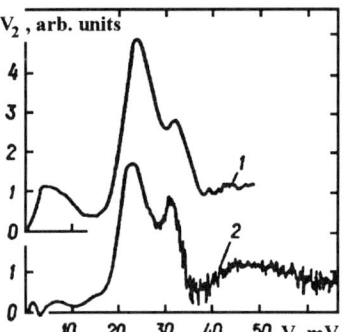

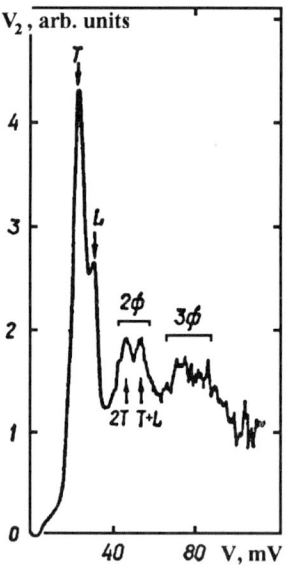

Figure 139 EPI point contact spectra for molybdenum:
1 - data [168] (a pressed needle-plane point contact, $R_0 = 25.6$ ohms, $V_{1,0} = 2100$ μV, $T = 1.5$ K);
2 - data [92] (a pressed point contact of the sliding type, $R_0 = 35$ ohms, $V_{1,0} = 815$ μV, $T = 1.95$ K.) The amplitude of the fluctuation decreases in the region of the two-phonon maximum at voltages of 48-50 mV.

Figure 140 An EPI point contact spectrum for molybdenum [92], containing features caused by two-phonon (2p) and three-phonon (3p) processes of the EPI. The maxima corresponding to scattering of electrons on transverse (T) and longitudinal (L) phonons, and the characteristic energies of two-phonon processes are indicated by arrows (a pressed point contact of the sliding type, $R_0 = 5.5$ ohms, $V_{1,0} = 630$ μV, $T = 4.2$ K).

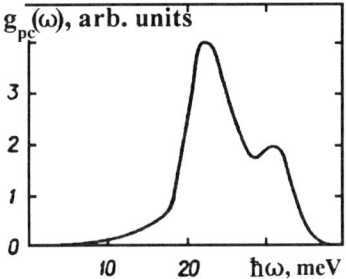

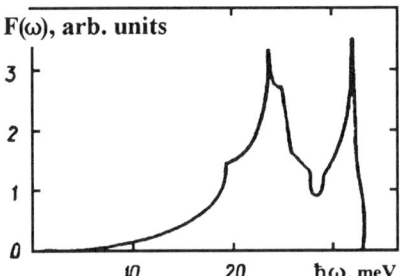

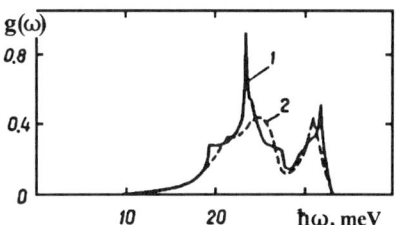

Figure 141 The EPI point contact function for molybdenum, reconstructed from the point contact spectrum in Figure 183.

Figure 142 The EPI thermodynamic function for molybdenum:
1 - calculation [276]; 2 - calculation of Glötzl, et. al. (cited in reference [169]).

Figure 143 The phonon density of states in molybdenum (calculation [277]).

Table 30 Values of the EPI point contact function for molybdenum.

$\hbar\omega$, meV	$\dfrac{g_{pc}(\omega)}{[g_{pc}(\omega)]_{max}}$	$\hbar\omega$, meV	$\dfrac{g_{pc}(\omega)}{[g_{pc}(\omega)]_{max}}$	$\hbar\omega$, meV	$\dfrac{g_{pc}(\omega)}{[g_{pc}(\omega)]_{max}}$
0	0	13.2	0.060	26.4	0.595
0.6	0.000	13.8	0.078	27	0.533
1.2	0.001	14.4	0.085	27.6	0.480
1.8	0.002	15	0.093	28.2	0.443
2.4	0.003	15.6	0.099	28.8	0.433
3	0.003	16.2	0.112	29.4	0.445
3.6	0.004	16.8	0.130	30	0.466
4.2	0.004	17.4	0.160	30.6	0.488
4.8	0.005	18	0.234	31.2	0.489
5.4	0.006	18.6	0.333	31.8	0.468
6	0.008	19.2	0.459	32.4	0.427
6.6	0.008	19.8	0.629	33	0.350
7.2	0.009	20.4	0.783	33.6	0.263
7.8	0.012	21	0.904	34.2	0.188
8.4	0.016	21.6	0.988	34.8	0.121
9	0.018	22.2	1	35.4	0.071
9.6	0.021	22.8	0.990	36	0.038
10.2	0.025	23.4	0.958	36.6	0.024
10.8	0.029	24	0.892	37.2	0.010
11.4	0.037	24.6	0.818	37.8	0.001
12	0.046	25.2	0.745	38.4	0.000
12.6	0.055	25.8	0.666		

2.21 Tungsten

Crystal lattice: BCC
References. Point contact spectroscopy: [9, 168, 169].
Phonon density of states: [55, 153, 197, 230, 249].

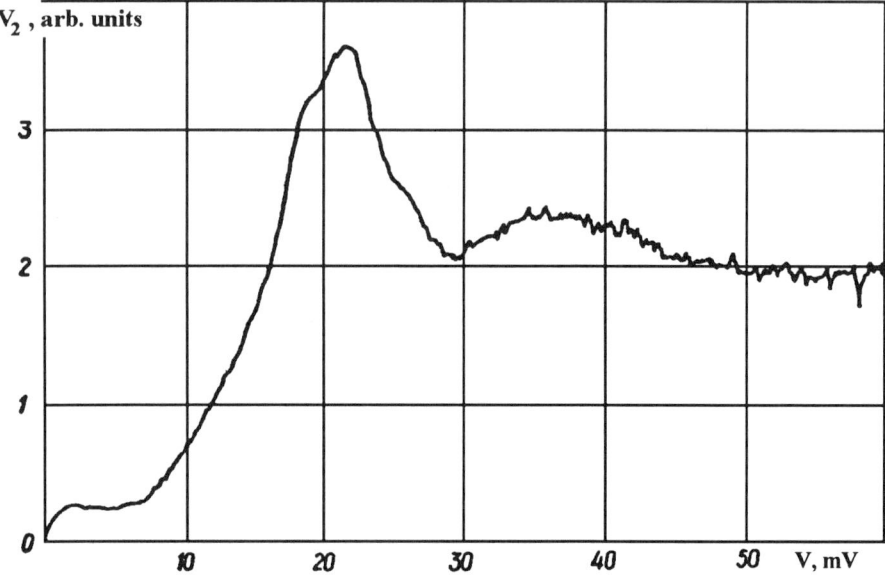

Figure 144 An EPI point contact spectrum for tungsten [92] (a pressed point contact of the sliding type, $R_0 = 21$ ohms, $V_{1,0} = 843$ μV, $V_{2max} = 0.69$ μV, T = 4.2 K).

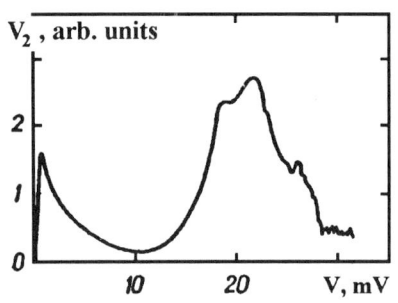 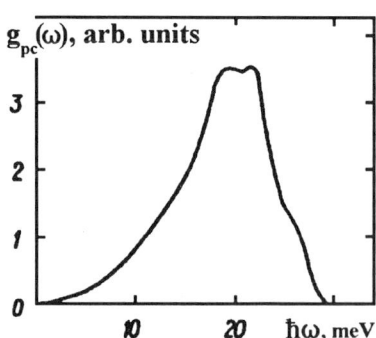

Figure 145 An EPI point contact spectrum for tungsten [168] (a pressed needle-plane point contact, $R_0 = 5.7$ ohms, $V_{1,0} = 830$ μV, T = 1.5 K).

Figure 146 The EPI point contact function for tungsten, reconstructed from the point contact spectrum in Figure 144.

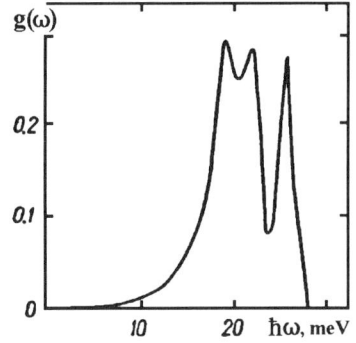
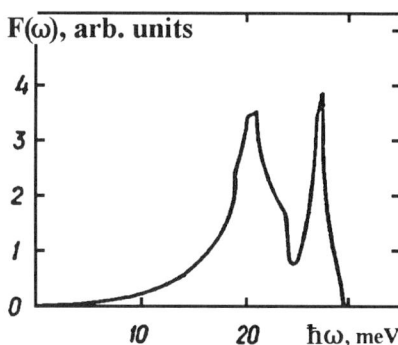

Figure 147 The EPI thermodynamic function for tungsten (calculation of Glötzl, et. al. (cited in reference [169])).

Figure 148 The phonon density of states in tungsten (calculation [197]).

Table 31 Values of the EPI point contact function for tungsten.

$\hbar\omega$, meV	$\dfrac{g_{pc}(\omega)}{[g_{pc}(\omega)]_{max}}$	$\hbar\omega$, meV	$\dfrac{g_{pc}(\omega)}{[g_{pc}(\omega)]_{max}}$	$\hbar\omega$, meV	$\dfrac{g_{pc}(\omega)}{[g_{pc}(\omega)]_{max}}$
0	0	10	0.247	20	0.985
0.5	0.003	10.5	0.275	20.5	0.987
1	0.010	11	0.303	21	1
1.5	0.014	11.5	0.330	21.5	0.985
2	0.020	12	0.352	22	0.948
2.5	0.027	12.5	0.387	22.5	0.872
3	0.031	13	0.415	23	0.762
3.5	0.038	13.5	0.448	23.5	0.642
4	0.045	14	0.484	24	0.556
4.5	0.051	14.5	0.522	24.5	0.473
5	0.060	15	0.562	25	0.410
5.5	0.073	15.5	0.600	25.5	0.351
6	0.081	16	0.648	26	0.306
6.5	0.097	16.5	0.703	26.5	0.256
7	0.105	17	0.780	27	0.203
7.5	0.123	17.5	0.871	27.5	0.146
8	0.144	18	0.943	28	0.095
8.5	0.165	18.5	0.991	28.5	0.054
9	0.192	19	0.994	29	0.021
9.5	0.225	19.5	0.993	29.5	0.000

2.22 Technetium

Crystal lattice: hcp
References. Point contact spectroscopy: [44]. Phonon density of states: [44].

Figure 149 An EPI point contact spectrum for technetium [44] (a pressed point contact of the sliding type, $R_0 = 1.5$ ohms, $V_1 = 1000$ μV, T = 4.2 K, B = 4.5 T).

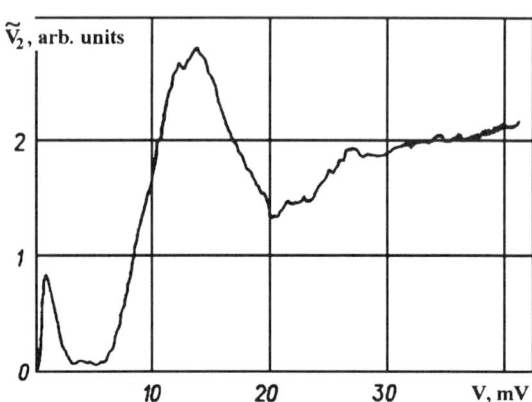

Table 32 Values of the EPI point contact function for technetium.

$\hbar\omega$, meV	$\dfrac{g_{pc}(\omega)}{[g_{pc}(\omega)]_{max}}$	$\hbar\omega$, meV	$\dfrac{g_{pc}(\omega)}{[g_{pc}(\omega)]_{max}}$	$\hbar\omega$, meV	$\dfrac{g_{pc}(\omega)}{[g_{pc}(\omega)]_{max}}$
0	0	10	0.671	20	0.115
0.5	0.005	10.5	0.747	20.5	0.078
1	0.011	11	0.840	21	0.076
1.5	0.012	11.5	0.913	21.5	0.079
2	0.014	12	0.962	22	0.079
2.5	0.015	12.5	0.988	22.5	0.081
3	0.020	13	1	23	0.089
3.5	0.022	13.5	0.990	23.5	0.092
4	0.024	14	0.958	24	0.097
4.5	0.026	14.5	0.889	24.5	0.100
5	0.031	15	0.810	25	0.109
5.5	0.044	15.5	0.722	25.5	0.117
6	0.067	16	0.641	26	0.132
6.5	0.107	16.5	0.555	26.5	0.144
7	0.170	17	0.477	27	0.142
7.5	0.237	17.5	0.412	27.5	0.118
8	0.353	18	0.347	28	0.090
8.5	0.435	18.5	0.287	28.5	0.061
9	0.516	19	0.215	29	0.038
9.5	0.598	19.5	0.169	29.5	0.000

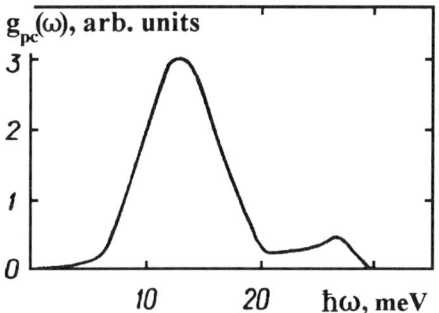

Figure 150 The EPI point contact function for technetium, reconstructed from the point contact spectrum for Figure 149.

Figure 151 The phonon density of states in technetium [44], obtained by thermal neutron scattering:
1 - T = 300 K, 2 - 150 K

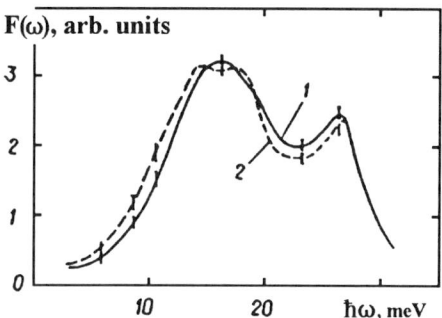

2.23 Rhenium

Crystal lattice: hcp
References. Point contact spectroscopy: [105, 106, 319].
Phonon density of states: [282].

Table 33 Values of the EPI point contact function for rhenium.

$\hbar\omega$, meV	$g_{pc}(\omega)/[g_{pc}(\omega)]_{max}$		$\hbar\omega$, meV	$g_{pc}(\omega)/[g_{pc}(\omega)]_{max}$	
	[0001] direction	(0001) plane		[0001] direction	(0001) plane
0	0	0	2	0.005	0.010
0.5	0.001	0.002	2.5	0.007	0.013
1	0.003	0.005	3	0.008	0.015
1.5	0.004	0.007	3.5	0.010	0.017

Table 33, contd.

$\hbar\omega$, meV	$g_{pc}(\omega)/[g_{pc}(\omega)]_{max}$		$\hbar\omega$, meV	$g_{pc}(\omega)/[g_{pc}(\omega)]_{max}$	
	[0001] direction	(0001) plane		[0001] direction	(0001) plane
4	0.011	0.020	17	0.550	0.718
4.5	0.012	0.022	17.5	0.451	0.622
5	0.014	0.025	18	0.383	0.512
5.5	0.015	0.027	18.5	0.341	0.385
6	0.016	0.034	19	0.332	0.330
6.5	0.018	0.036	19.5	0.311	0.297
7	0.019	0.041	20	0.300	0.276
7.5	0.020	0.042	20.5	0.281	0.334
8	0.022	0.047	21	0.272	0.349
8.5	0.028	0.054	21.5	0.256	0.296
9	0.048	0.067	22	0.239	0.239
9.5	0.087	0.083	22.5	0.230	0.235
10	0.128	0.098	23	0.236	0.277
10.5	0.166	0.120	23.5	0.238	0.330
11	0.242	0.146	24	0.236	0.440
11.5	0.360	0.184	24.5	0.232	0.534
12	0.538	0.261	25	0.217	0.514
12.5	0.703	0.394	25.5	0.194	0.470
13	0.890	0.559	26	0.156	0.242
13.5	0.963	0.723	26.5	0.126	0.139
14	0.992	0.860	27	0.092	0.072
14.5	0.859	0.929	27.5	0.065	0.011
15	0.705	0.996	28	0.025	0.000
15.5	0.682	0.982	28.5	0.005	
16	0.660	0.909	29	0.000	
16.5	0.625	0.812			

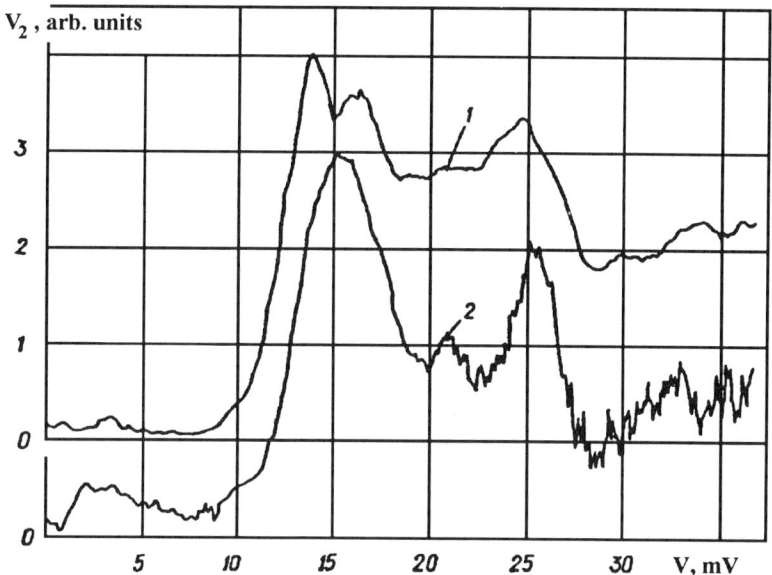

Figure 152 EPI point contact spectra for rhenium [105]:
1 - [0001] direction, 2 - (0001) plane (pressed point contacts, 1 - $R_0 = 20.7$ ohms, $V_{1,0} = 210$ μV, $V_{2max} = 0.89$ μV, T = 2.2 K; 2 - $R_0 = 23.1$ ohms, $V_{1,0} = 250$ μV, $V_{2max} = 0.14$ μV, T = 2.4 K).

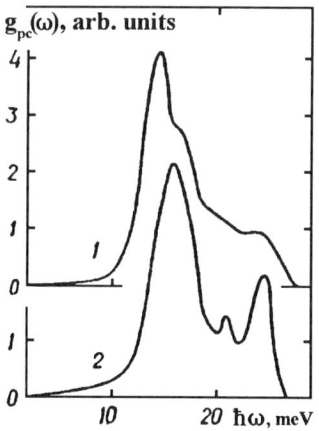
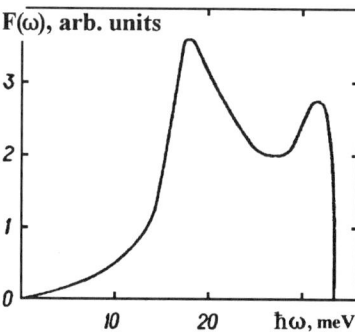

Figure 153 The EPI point contact function for rhenium (reconstructed from the corresponding point contact spectra in Figure 152):
1 - [0001] direction; 2 - (0001) plane.

Figure 154 The phonon density of states in rhenium (calculation [282]).

2.24 Iron

Crystal lattice: α-phase - BCC
References. Point contact spectroscopy: [22, 72, 326].
Phonon density of states: [24, 25, 28, 98, 153, 154, 230, 249, 256].

Figure 155
An EPI point contact spectrum for α-iron [72] (a pressed needle-plane point contact, $R_0 = 3.8$ ohms, $V_{1,0} = 450$ μV, $V_{2max} = 1.24$ μV, T = 1.5 K).

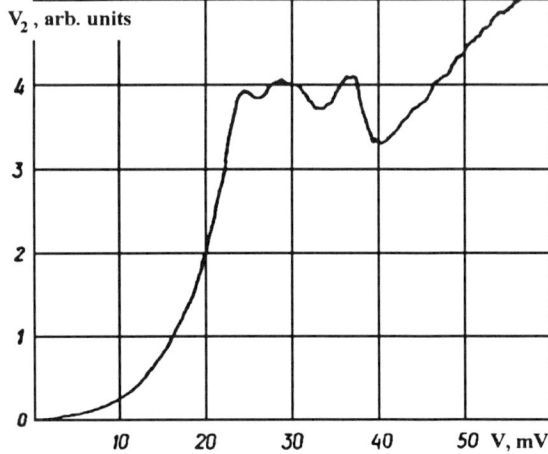

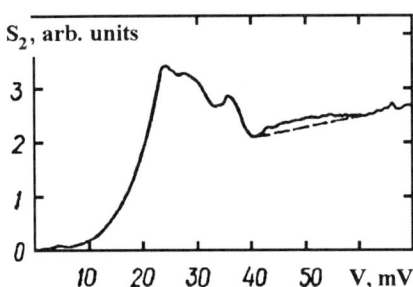

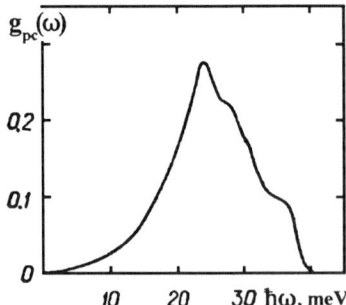

Figure 156 An EPI point contact spectrum for α-iron, (the same point contact as that in Figure 155, $V_1 = 450$ μV).

Figure 157 The EPI point contact function for α-iron, reconstructed from the point contact spectrum in Figure 155 ($\lambda_{pc} = 0.37$).

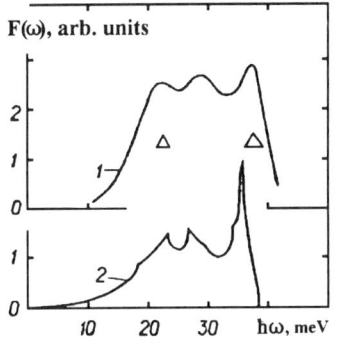

Figure 158 The phonon density of states in α-iron:
1 - data [28] obtained by thermal neutron scattering (the triangles indicate the resolution of the experimental setup);
2 - calculation [256].

Table 34 Values of the EPI point contact function for α-iron.

$\hbar\omega$, meV	$g_{pc}(\omega)$	$\hbar\omega$, meV	$g_{pc}(\omega)$	$\hbar\omega$, meV	$g_{pc}(\omega)$
0	0	13.8	0.058	27.6	0.222
0.6	0.001	14.4	0.065	28.2	0.214
1.2	0.001	15	0.070	28.8	0.205
1.8	0.002	15.6	0.079	29.4	0.192
2.4	0.002	16.2	0.090	30	0.180
3	0.003	16.8	0.099	30.6	0.171
3.6	0.004	17.4	0.108	31.2	0.156
4.2	0.004	18	0.119	31.8	0.142
4.8	0.006	18.6	0.131	32.4	0.128
5.4	0.008	19.2	0.150	33	0.117
6	0.010	19.8	0.164	33.6	0.108
6.6	0.010	20.4	0.182	34.2	0.103
7.2	0.012	21	0.197	34.8	0.099
7.8	0.013	21.6	0.216	35.4	0.097
8.4	0.015	22.2	0.235	36	0.093
9	0.019	22.8	0.257	36.6	0.088
9.6	0.022	23.4	0.267	37.2	0.078
10.2	0.025	24	0.277	37.8	0.057
10.8	0.029	24.6	0.266	38.4	0.036
11.4	0.033	25.2	0.255	39	0.019
12	0.037	25.8	0.242	39.6	0.008
12.6	0.043	26.4	0.231	40.2	0.000
13.2	0.051	27	0.225		

2.25 Cobalt

Crystal lattice: α-phase - hcp
References. Point contact spectroscopy: [29, 326].
Phonon density of states: [204, 283, 295, 296, 310].

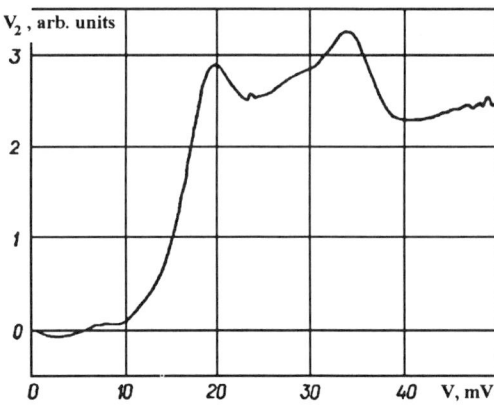

Figure 159 An EPI point contact spectrum for α-cobalt [29] (a pressed point contact of the sliding type, $R_0 = 32$ ohms, $V_{1,0} = 855$ μV, $V_{2max} = 1.67$ μV, $T = 3$ K).

Table 35 Values of the EPI point contact function for α-cobalt.

ℏω, meV	$g_{pc}(\omega)$	ℏω, meV	$g_{pc}(\omega)$	ℏω, meV	$g_{pc}(\omega)$
0	0	14.5	0.096	28.5	0.156
0.5	0.000	15	0.117	29	0.154
1	0.000	15.5	0.145	29.5	0.151
1.5	0.000	16	0.175	30	0.149
2	0.000	16.5	0.202	30.5	0.146
2.5	0.000	17	0.228	31	0.146
3	0.001	17.5	0.256	31.5	0.148
3.5	0.001	18	0.283	32	0.152
4	0.002	18.5	0.303	32.5	0.155
4.5	0.002	19	0.313	33	0.156
5	0.002	19.5	0.310	33.5	0.155
5.5	0.002	20	0.295	34	0.151
6	0.003	20.5	0.273	34.5	0.141
6.5	0.003	21	0.249	35	0.128
7	0.003	21.5	0.230	35.5	0.111
7.5	0.004	22	0.214	36	0.092
8	0.004	22.5	0.199	36.5	0.072
8.5	0.005	23	0.187	37	0.054
9	0.006	23.5	0.179	37.5	0.038
9.5	0.006	24	0.174	38	0.026
10	0.007	24.5	0.170	38.5	0.018
10.5	0.013	25	0.167	39	0.013
11	0.021	25.5	0.165	39.5	0.009
11.5	0.028	26	0.164	40	0.006
12	0.037	26.5	0.163	40.5	0.004
12.5	0.046	27	0.161	41	0.003
13	0.055	27.5	0.160	41.5	0.002
13.5	0.066	28	0.158	42	0.000
14	0.080				

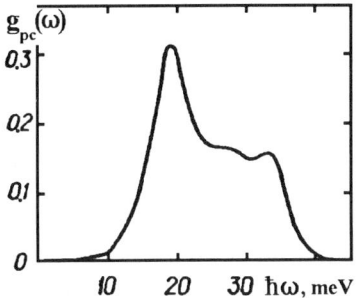 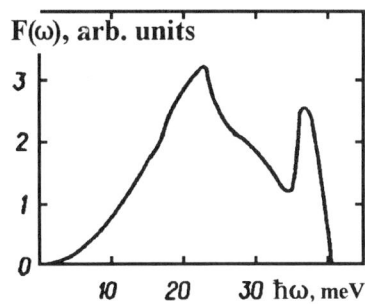

Figure 160 The EPI point contact function for α-cobalt, reconstructed from the point contact spectrum in Figure 159 ($\lambda_{pc} = 0.39$).

Figure 161 The phonon density of states in α-cobalt (calculation [283]).

2.26 Nickel

Crystal lattice: FCC
References. Point contact spectroscopy: [22, 72, 243, 326].
Phonon density of states: [26, 53, 102, 121, 157, 200, 229, 266, 278, 300].

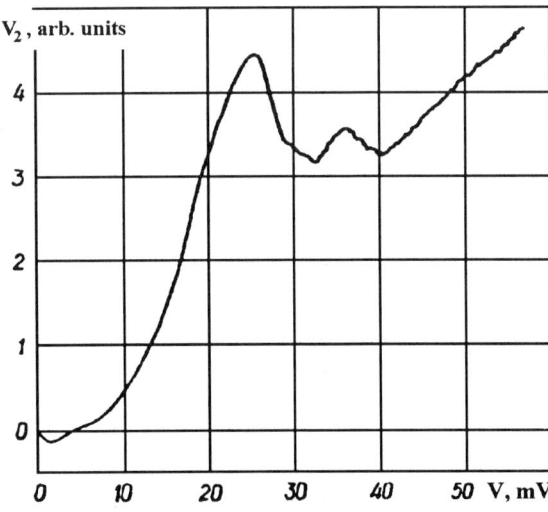

Figure 162 An EPI point contact spectrum for nickel [72] (a pressed needle-plane point contact, $R_0 = 2.5$ ohms, $V_{1,0} = 750$ μV, $V_{2max} = 1.82$ μV, $T = 2.5$ K).

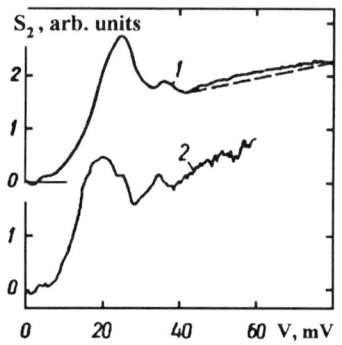 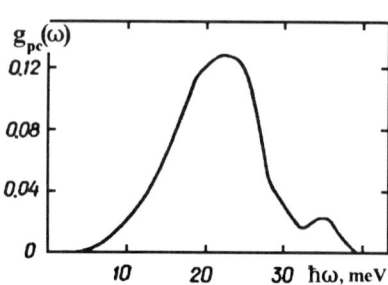

Figure 163 EPI point contact spectra for nickel [72]: 1 - the same point contact as that in Figure 162, $V_1 = 750$ μV; 2 - an atypical spectrum of a pressed needle-plane point contact between monocrystalline electrodes with an arbitrary orientation, displaying possible anisotropy of the EPI ($R_0 = 9.3$ ohms, $V_1 = 800$ μV, T = 4.2 K).

Figure 164 The EPI point contact function for nickel, reconstructed from the point contact spectrum in Figure 162 ($\lambda_{pc} = 0.19$).

Table 36 Values of the EPI point contact function for nickel.

$\hbar\omega$, meV	$g_{pc}(\omega)$	$\hbar\omega$, meV	$g_{pc}(\omega)$	$\hbar\omega$, meV	$g_{pc}(\omega)$
0	0	13.5	0.048	27	0.085
0.5	0.000	14	0.053	27.5	0.072
1	0.000	14.5	0.057	28	0.059
1.5	0.000	15	0.064	28.5	0.048
2	0.000	15.5	0.070	29	0.042
2.5	0.000	16	0.076	29.5	0.038
3	0.000	16.5	0.084	30	0.034
3.5	0.000	17	0.090	30.5	0.030
4	0.000	17.5	0.097	31	0.026
4.5	0.001	18	0.104	31.5	0.023
5	0.001	18.5	0.113	32	0.020
5.5	0.002	19	0.117	32.5	0.017
6	0.003	19.5	0.121	33	0.018
6.5	0.004	20	0.124	33.5	0.020
7	0.006	20.5	0.127	34	0.022
7.5	0.008	21	0.129	34.5	0.024
8	0.009	21.5	0.131	35	0.026
8.5	0.012	22	0.132	35.5	0.025
9	0.015	22.5	0.133	36	0.024
9.5	0.018	23	0.133	36.5	0.021
10	0.021	23.5	0.131	37	0.017
10.5	0.024	24	0.130	37.5	0.014
11	0.027	24.5	0.127	38	0.010
11.5	0.031	25	0.124	38.5	0.006
12	0.035	25.5	0.118	39	0.004
12.5	0.039	26	0.110	39.5	0.002
13	0.043	26.5	0.098	40	0.000

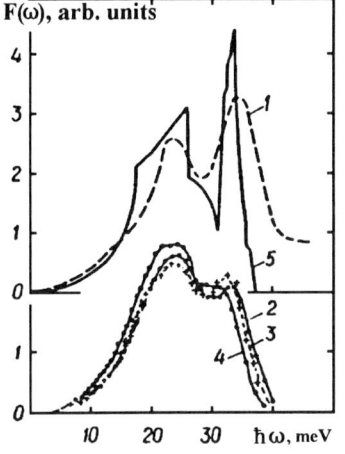

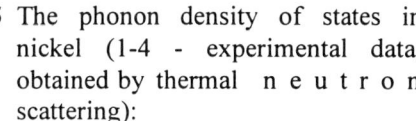

Figure 165 The phonon density of states in nickel (1-4 - experimental data, obtained by thermal n e u t r o n scattering):
1 - results of [121];
2-4 - results of [26] for samples with different isotopic compositions (2 - natural nickel, 3 - from incoherent scattering on Ni, 4 - ^{62}Ni isotope), 5 - calculation [157].

2.27 Palladium

Crystal lattice: FCC
References. Point contact spectroscopy: [168, 169]. EPI functions: [276]. Phonon density of states: [255, 267, 269].

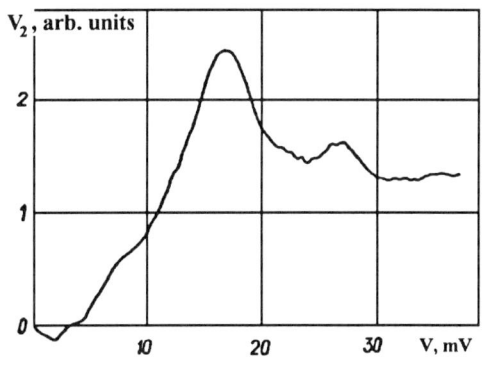

Figure 166 An EPI point contact spectrum for palladium [168] (a pressed needle-plane point contact, R_0 = 20.5 ohms, $V_{1,0}$ = 2000 μV).

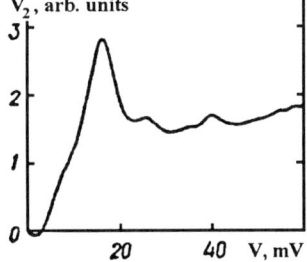

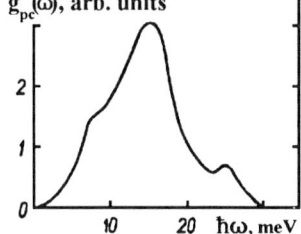

Figure 167 An EPI point contact spectrum for palladium [168] (a pressed needle-plane point contact, R_0 = 6.8 ohms, $V_{1,0}$ = 2100 μV, T = 1.5 K).

Figure 168 The EPI point contact function for palladium [172].

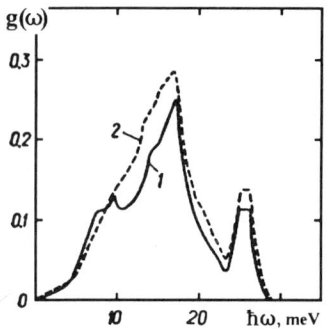
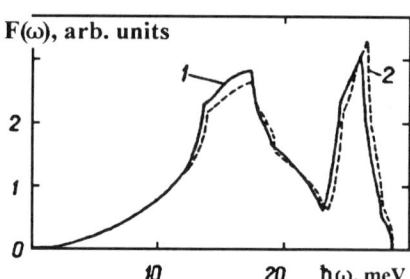

Figure 169 The EPI thermodynamic function for palladium:
1 - calculation [276 ; 2 - calculation of Glötzl, et. al. (cited in reference [169]).

Figure 170 The phonon density of states in palladium (calculation [255]):
1 - T = 296 K ; 2 - 120 K.

2.28 Osmium

Crystal lattice: hcp
References. Point contact spectroscopy: [111].

Table 37 Values of the EPI point contact function for osmium.

$\hbar\omega$, meV	$g_{pc}(\omega)$	$\hbar\omega$, meV	$g_{pc}(\omega)$	$\hbar\omega$, meV	$g_{pc}(\omega)$
0	0	10	0.023	20	0.522
0.5	0.000	10.5	0.028	20.5	0.447
1	0.000	11	0.032	21	0.391
1.5	0.000	11.5	0.036	21.5	0.352
2	0.000	12	0.041	22	0.347
2.5	0.000	12.5	0.045	22.5	0.338
3	0.000	13	0.059	23	0.312
3.5	0.000	13.5	0.073	23.5	0.271
4	0.000	14	0.100	24	0.226
4.5	0.000	14.5	0.132	24.5	0.185
5	0.000	15	0.187	25	0.165
5.5	0.000	15.5	0.258	25.5	0.172
6	0.000	16	0.340	26	0.178
6.5	0.002	16.5	0.405	26.5	0.171
7	0.005	17	0.465	27	0.137
7.5	0.007	17.5	0.547	27.5	0.079
8	0.009	18	0.639	28	0.044
8.5	0.012	18.5	0.698	28.5	0.022
9	0.016	19	0.680	29	0.010
9.5	0.018	19.5	0.602	29.5	0.000

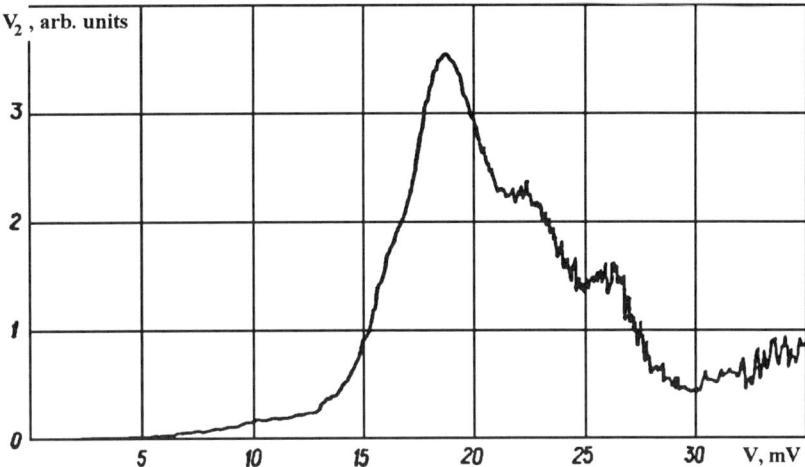

Figure 171 An EPI point contact spectrum for osmium [111] (a pressed point contact of the sliding type, $R_0 = 5.5$ ohms, $V_{1,0} = 372$ μV, $V_{2max} = 1.12$ μV, T = 1.6 K).

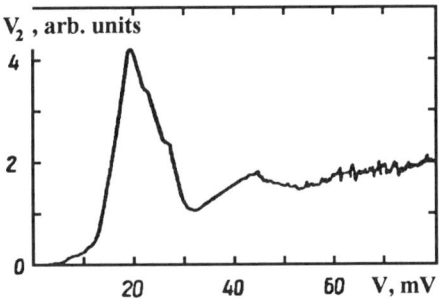

Figure 172 An EPI point contact spectrum for osmium [111]:
(a pressed point contact of the sliding type, $R_0 = 5.3$ ohms, $V_{1,0} = 501$ μV, T = 1.5 K).

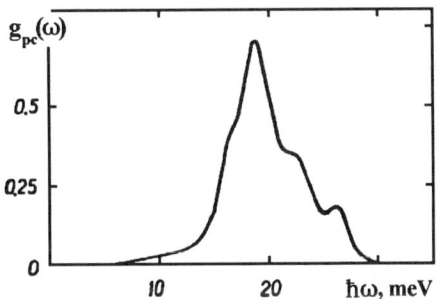

Figure 173 The EPI point contact function for osmium, reconstructed from the point contact spectrum in Figure 171 ($\lambda_{pc} = 0.52$).

2.29 Gadolinium

Crystal lattice: hcp
References. Point contact spectroscopy: [2,7]. Phonon density of states: [281].

Figure 174 A point contact spectrum for gadolinium [2] (a pressed point contact of the sliding type, $R_0 = 37$ ohms, $V_{1,0} = 800$ μV, T = 4.2 K).

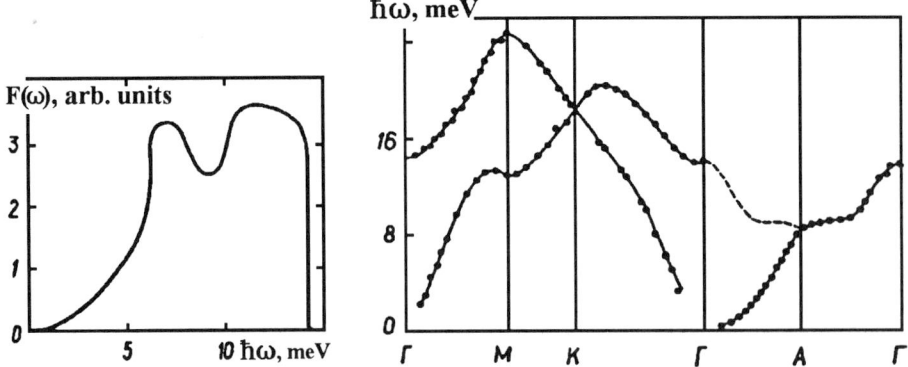

Figure 175 The phonon density of states in gadolinium (calculation [281]).

Figure 176 Dispersion curves for magnons in gadolinium at T = 78 K [234].

2.30 Terbium

Crystal lattice: hcp

References. Point contact spectroscopy: [2,7, 239]. Phonon density of states: [213]. Magnon density of states : [257].

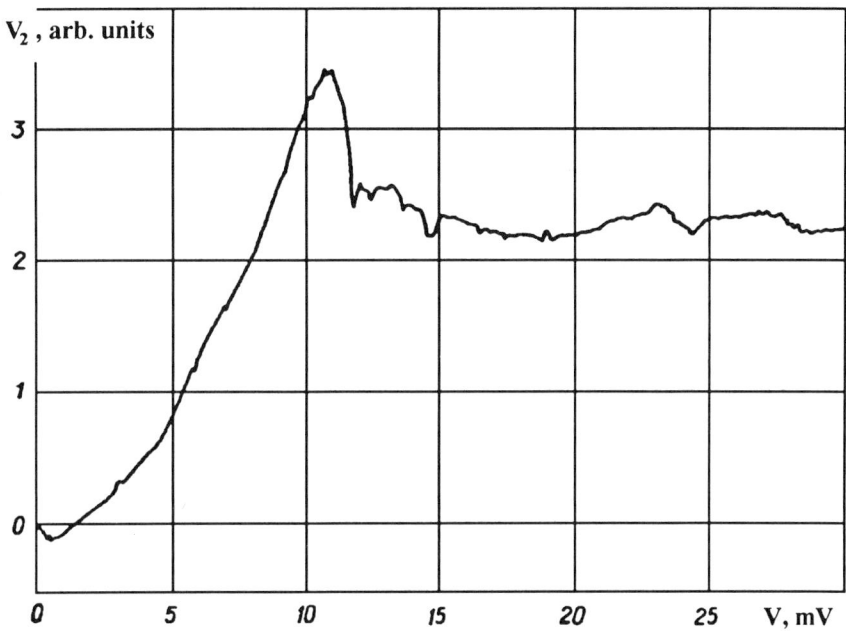

Figure 177 A point contact spectrum for terbium [2] (a pressed point contact of the sliding type, $R_0 = 4$ ohms, $V_{1,0} = 600$ μV, T = 1.9 K).

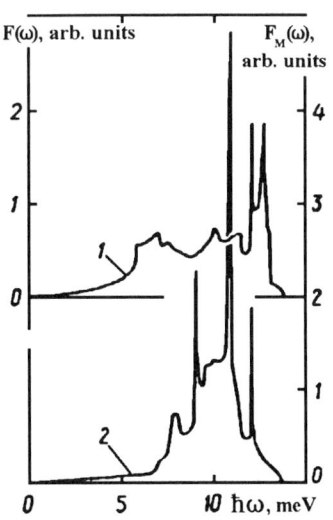

Figure 178 The density of states of elementary Bose-type excitations in terbium:
1 - phonon density of states (calculation [213]);
2 - magnon density of states (calculation [257]).

2.31 Holmium

Crystal lattice: hcp
References. Point contact spectroscopy: [2,7]. Phonon density of states: [261].

Figure 179
Point contact spectra for holmium [2]: (a pressed point contact of the sliding type, $V_{1,0} = 800$ μV, T = 1.7 K); 1 - R_0 = 50 ohms; 2 - 40 ohms.

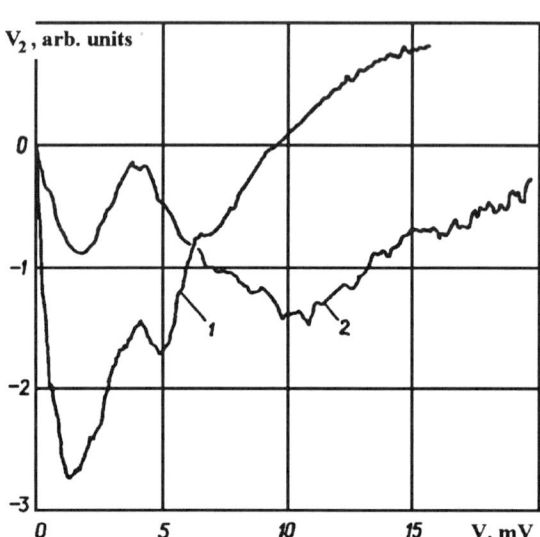

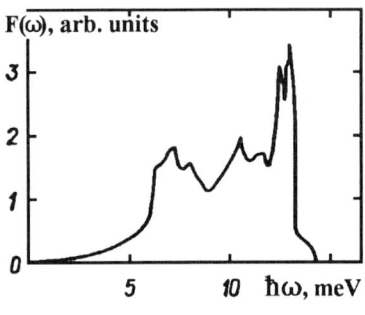

Figure 180 The phonon density of states in holmium (calculation [261]).

Figure 181
Dispersion curves of magnons in holmium at T = 4.6 K [308]:
1 - [0001] direction;
2 - [11$\bar{2}$0] direction.

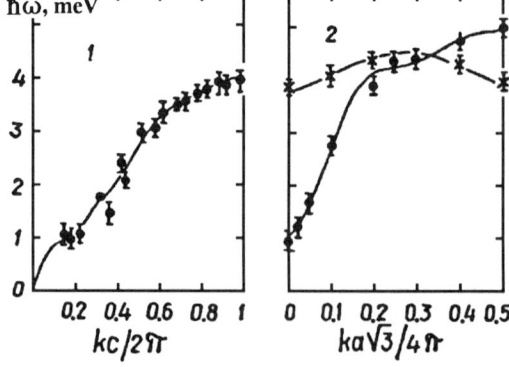

Table 38   EPI Parameters (λ_{pc}, λ)

Metal	Crystal lattice	Lattice constants	$\varkappa$	v_F, (10^6 m/s)	λ_{pc}	λ
Sodium	BCC	a = 4.23	1	1.07	0.068	0.16±0.03
Potassium	BCC	a = 5.225	1	0.86	0.11	0.13±0.03
Copper	FCC	a = 3.603	1	1.57	0.24	0.15±0.03
Silver	FCC	a = 4.061	1	1.40	0.1	0.13±0.04
Gold	FCC	a = 4.080	1	1.39	0.095	0.17±0.05
Beryllium	hcp	a = 2.278 c = 3.579	4	2.25	0.24	0.24±0.05
Magnesium	hcp	a = 3.209 c = 5.210	4	1.59	0.28	0.35±0.05
Zinc	hcp	a = 2.665 c = 4.947	4	1.82	0.57	0.37±0.05
Cadmium	hcp	a = 2.97 c = 5.55	4	1.63	0.44	0.40±0.05
Aluminum	FCC	a = 4.032	3	2.03	0.45	0.43±0.05
α-Gallium	rhombohedral	a = 4.516 b = 4.490 c = 7.663	12	1.92	0.31	-
Indium	Face-centered tetragonal	a = 4.556 c = 4.934	6	1.75	0.85	0.80±0.05
α-Thallium	hcp	a = 3.438 c = 5.478	6	1.70	0.78	0.80±0.05
β-Tin	Body-centered tetragonal	a = 5.831 c = 3.181	4	1.89	0.71	0.72±0.05
Lead	FCC	a = 4.095	4	1.84	1.7	1.55±0.05
Niobium	BCC	a = 3.300	1	1.37	0.81	1.0±0.3
α-Iron	BCC	a = 2.861	2	1.99	0.37	-
α-Cobalt	hcp	a = 2.502 c = 4.059	2	2.03	0.39	-
Nickel	FCC	a = 3.514	2	2.04	0.19	-
Osmium	hcp	a = 2.735 c = 4.318	4	1.87	0.52	0.4±0.1

Table 39   A few fundamental constants

Constant	symbol	numerical value
Planck's constant	$\hbar$	1.05459×10^{-34} J·s
electron charge	e	1.60219×10^{-19} C
free electron mass	m	9.1095×10^{-31} kg
Boltzmann constant	k	1.3806×10^{-23} J/K (0.8617×10^{-4} eV/K)
electron volt	eV	1.602×10^{-19} J 1.1604×10^{4} K 2.41796×10^{14} Hz 8.067×10^{3} cm^{-1}

References

Abbreviations

DAN - Doklady Akademii Nauk SSSR (Soviet Physics Doklady)

JTP - Zhurnal Tekhnicheskoi Fiziki (Soviet Physics - Technical Physics)

JETP - Zhurnal Eksperimental'noi Teoretischeskoi Fiziki (Soviet Physics-JETP)

PTE - Pribory i Tekhnika Eksperimenta (Instruments and Experimental Techniques (USSR))

UFJ - Ukainskii Fizicheskii Zhurnal (Ukrainian Physics Journal)

UFN - Uspekhi Fizicheskikh Nauk (Soviet Physics-Uspekhi)

FMM - Fizika Metallov i Metallovedenie (Physics of Metals and Metallography(USSR))

FNT - Fizika Nizkikh Temperatur (Soviet Journal of Low Temperature Physics)

FTINT - Physical-Technical Low Temperature Institute

FTP - Fizika i Tekhnika Poluprovodnikov (Soviet Physics-Semiconductors)

FTT - Fizika Tverdogo Tela (Soviet Physics-Solid State)

1. Akimenko, A. I., Verkin, A. B., Ponomarenko, N. M., Yanson, I. K., Dependence of the Resistance of a Cu-Cu Microcontact at Temperatures in the Range 1.5-100 K. - FNT, 1981, **7**, no.8, pp. 1076-1080.

2. Akimenko, A. I., Verkin, A. B., Ponomaremko, N. M., Yanson, I. K., Microcontact Spectroscopy of Magnons in Rare-Earth Metals - FNT, 1982, **8**, no.10, pp. 1084-1094.

3. Akimenko, A. I., Verkin, A. B., Ponomarenko, N. M., Yanson, I. K., Microcontact Spectroscopy of Copper: Temperature Measurement. Dependence of the MC Spectra on Contact Parameters. - FNT, 1982, **8**, no.3, pp. 260-275.

4. Akimenko, A. I., Verkin, A. B., Yanson, I. K.,Anisotropy of the Microcontact Noise Spectra of Copper - FNT, 1984, **10**, no.11, pp. 1159-1165.

5. Akimenko, A. I., Ponomarenko, N. M., Yanson, I. K., et. al., Microcontact Spectroscopy of the Intercrystalline Region in $PrNi_5$ - FTT, 1984, **26**, issue 8, pp. 2264-2272.

6. Akimenko, A. I., Yanson, I. K., Dependence of the Resistance of Point Microcontacts on Controlled Variation of Their Size - Letter in JTP, 1978, **4**, issue 14, pp. 856-860.

7. Akimenko, A. I., Yanson, I. K., Microcontact Spectroscopy of Magnons in Metals. - Letter in JETP, 1980, **31**, issue 4, pp. 209-213.

8. Akimenko, A. I., Yanson, I. K., Microcontact Spectra of $PrNi_5$ in the Thermal Regime : Comparison with Theory. - FNT, 1984, **10**, no.8, pp. 889-892.

9. Artemenko, S. N., Volkov, A. F., Zaitzev, A. V., Theory of the Nonstationary Josephson Effect in Short Superconducting Contacts - JETP, 1979, **76**, issue 5, pp. 1816-1833.

10. Artemenko, S. N., Volkov, A. F., A Weak Link Between Conductors with a Charge Density Wave. - JETP, 1984, **87**, issue 2, pp. 691-701.

11. Aslamazov, L. G., Larkin, A. I., The Josephson Effect in Superconducting Point Contacts. - Letter in JETP, 1969, **9**, issue 2, pp. 150-154.

12. Batrak, A. G., Lavrent'ev, F. F., Nikoforenko, V. N., Yanson, I. K.,Influence of Dislocations on the Microcontact Spectroscopy of Zinc Monocrystals. - FNT, 1980, **6**, no. 11, pp. 1446-1451.

13. Bartak, A. G., Yanson, I. K., Features of the Microcontact Spectra of Cu-Zn Heterocontacts - FNT, 1979, **5**, no.12, pp. 1404-1408.

14. Bogomolov, V. N., Klushin, N. A., Okuneva, N. M., et. al. Investigation of the Phonon Spectrum of Gallium in Porous Glass - FTT, 1971, **13**, issue 5, pp. 1499-1501.

15. Born, M., Huan Kun, The Dynamic Theory of Crystal Lattices - M.: Pub. House of Foreign Lit., 1958, 488 pp..

16. Brandt, N. B., Chudinov, S. M., Experimental Methods of Investigating the Energy Spectra of Electrons and Phonons in Metals - M.: Pub. House of MSU, 1983, 403 pp.

17. Bredov, M. M., Kotov, B. A., Okuneva, N. M., et. al. On the Possibility of Measurement of the Spectra of Thermal Oscillations by Means of Coherent Inelastic Neutron Scattering on Polycrystals - FTT, 1967, **9**, issue 1, pp. 287-293.

18. Brovman, E. G., Kagan, Y., On the Phonon Spectrum of the Lattice of White Tin. - FTT, 1966, **8**, issue 5, pp. 1042-1416.

19. Bulat, I. A., The Phonon Spectrum of Beryllium from Data Measured Using Slow Neutrons - FTT, 1979, **21**, issue 4, pp. 1001-1005.

20. Bulat, I. A., Makovetskii, G. I., The Phonon Spectrum and Thermodynamic Functions of Magnesium - FMM, 1978, **45**, issue 3, pp. 636-640.

21. Vedeneev, S. I., Golovashkin, A. I., Koshelkora, E. V., Motulevich, G. P., Program for Processing of the Results of Tunneling Investigations and Reconstruction of the Phonon Spectra of Superconductors (program of McMillan - Rowell - Kuban) - M., 1975, 41 pp. -(Preprint/ Phys. Institute AS - USSR ; No. 57).

22. Verkin, B. I., Yanson, I. K., Kulik, I. O., et. al. Modulation of the Temperature of Elementary Excitation Spectroscopy in Ferromagnets With the Aid of Microcontacts - Izv. Akad. Nauk. USSR, Ser. Phys., 1980, **44**, no.7, pp. 1330-1338.

23. Galkin, A. A., D'yachenko, A. I., Svistunov, V. M. Determination of the Energy Gap Parameter and Electron- Phonon Interaction Function From Tunneling Data - JETP, 1974, **66**, issue 6, pp. 2262-2268.

24. Gorbachev, B. I., Ivanitskii, P. G., Korotenko, V. T., Pasechnik, M. V., Studies of Inelastic Thermal Neutron Scattering on Samples of Iron in the Temperature Interval 24 - 294 °C. - UFJ, 1973, **18**, No.9, pp. 1528-1534.

25. Gorbachev, B. I., Ivanitskii, P. G., Korotenko, V. T., Pasechnik, M. V., Investigation of the Inelastic Scattering of Thermal Neutrons in Polycrystalline Samples of Iron. - UFN, 1973, **18**, No.8, pp. 1384-1389.

26. Gorbachev, B. I., Ivanitskii, P. G., Korotenko, V. T., Pasechnik, M. V., Investigation of Thermal Neutron Scattering in Nickel Samples with Different Isotopic Composition - UFJ, 1973, **18**, No.4, pp. 558 - 563.

27. Gorbachev, B. I., Ivanitskii, P. G., Korotenko, V. T., Pasechnik, M. V., The Phonon Spectrum of the Copper Lattice - UFJ, 1973, **18**, No.8, pp. 1390 - 1392.

28. Gorbachev, B. I., Morozov, S. I., Parfenov, V. A., Pasechnik, M. V., The Phonon Spectrum of an α-Fe Crystal - FTT, 1976, **18**, issue 12, pp. 3702 - 3704.

29. Gribov, N. N., Microcontact Spectra of the Electron- Phonon Interaction in Cobalt - FNT, 1984, **10**, No.3, pp. 324 - 326.

30. Gurevich, I. I., Tarasov, L. V., The Physics of Low Energy Neutrons - M.: Science, 1965, 607 pp.

31. Jeklevik, R. K., Lamb, J., Molecular Excitations in Barriers II. - in the book: Tunneling Phenomena in the Solid State. M.: World, 1973, pp. 233 - 243.

32. Dobretsov, A. I., Tulina, N. A., Tuflin, Y. A., A Sensitive Bridge Circuit for Observing Volt-amperage Characteristics - PTE, 1982, No.3, pp. 136 - 138.

33. Eremeev, I. P., Sadukov, I. P., Chernishov, A. A. - The Phonon Density of States in hexagonal metals of group II - FTT, 1976, **18**, issue 6, pp. 1652 - 1660.

34. Eremeev, I. P., Saidkov, I. P., Chernishov, A. A., Experimental Determination of the Phonon Density of States in Crystals by Coherent Scattering. - FTT, 1973, **15**, issue 7, pp. 1953 - 1962.

35. Eremeev, I. P., Chernushov, A. A., Sadikov, I. P., The Phonon Density of States in Cadmium. - Letter in JETP, 1973, **18**, issue 5, pp. 302 - 305.

36. Zhernov, A. P., Kulagin, V. D., Kulagina, T. N., On the Nature of Distinguishing the Electron-phonon Interaction Microcontact Function of Single Valence Metals. - FTT, 1984, **26**, issue 2, pp. 587 - 589.

37. Zhernov, A. P., Kulagina, T. N., Naidyuk, Yu. G., Yanson, I. K., Determination of the Electron-phonon Interaction Function in Sodium, Potassium and Aluminum from 'First Principles'. - M., 1982 - 12 pp. - (Preprint/ Inst. of Atomic Energy : No. 3637/10).

38. Zhernov, A. P., Naidyuk, Yu. G., Yanson, I. K., Kulagina, T. N., Determination of the Electron-Phonon Interaction Microcontact Function in Sodium and Potassium by the Pseudopotential Method. - FNT, 1982, **8**, No. 7, pp. 713 - 717.

39. Zavaritskii, N. V., Itskevitch, E. S., Voronovskii, A. N., Variation of the Vibrational Spectrum of the Lattice and the Electron-Phonon Interaction in Superconductors Under Pressure. - JETP, 1971, **60**, issue 4, pp. 1408 - 1417.

40. Zavaritskii, N. V., On the Increase of the Critical Temperature of Superconductors, Condensation at Low Temperatures. - JETP, 1969, **57**, issue 3, pp. 752 - 762.

41. Zavaritskii, I. V., The Electron-Phonon Interaction and Characteristics of Electrons in Metals. - UFN, 1972, **108**, issue 2, pp. 241 - 272.

42. Zaitsev, A. V., A Theory of Clean Shorted Microbridges, S-c-S and S-c-N. - JETP, 1980, **78**, issue 1, pp. 221 -233.

43. Zaitsev, A. V., Quasiclassical Equation Theory of Superconductivity for Contacting Metals and the Relationship of Microcontacts with Narrowings. - JETP, 1984, **86**, issue 5, pp. 1742 - 1757.

44. Zakharov, A. A., Zemlyanov, M. G., Mikheyeva, M. N., et. al. Investigation of the Electron-Phonon Interaction (EPI) and Phonon Density of States in Technetium by the Method of Microcontact Spectroscopy and Incoherent Neutron Scattering. - In the book : 23-e All Union Conference on Low Temperature Physics (Tallin, 23 - 25 Oct. 1984): Inst. Chem. and Biol. Physics - Akad. Nauk ESSR, 1984, ch. 2, pp. 190 - 191.

45. Zemlyanov, M. G., Chernoplekov, N. A., Arrangement for Investigating the Dynamics of a Material in the Condensed State Utilizing Inelastic Scattering of Thermal Neutrons. - PTE, 1962, No.5, pp. 40 - 47.

46. Itsikovich, I. F., Shekhter, R. I., Quantum Theory of the Nonlinear Electrical Conduction of Semiconductor Microcontacts - FNT, 1985, **11**, No. 4, pp. 373 - 388.

47. Itsikovich, I. F., Shekhter, R. I., Theory of Multiphonon Spectra in Semiconductor Microcontacts. - FNT, 1984, **10**, No.4, pp. 437 - 442.

48. Kvatchev, A. P., Svistunov, V. M., Chubar, V. A., Measurement of the Characteristics of Tunnel Junctions Under Pressure. - Physics and Techniques of High Pressures, 1981, issue 4, pp. 70 - 76.

49. Kovalenko, V. S., Metallographic Reactions. Handbook - Metallurgia (Moscow), 1981. - 120 pp.

50. Kozub, V. I., On Low Temperature Characteristics of Point and Bridge Microcontacts of Normal Metals Having Amorphous Regions. - FTT, 1984, **26**, issue 7, pp. 1955- 1962.

51. Korn, G., Korn, T., Handbook of Mathematics. - M.: Science, 1968, 720 pp.

52. Korshunov, V. A., Determination of the Phonon Density of States with Thermodynamic Functions of the Crystal: Gold. - Proc. of the Inst. of Physics, 1976, No.3, pp. 145-146.

53. Korshunov, V. A., Determination of the Phonon Density of States with Thermodynamic Functions of the Crystal: Nickel, Palladium, Platinum. - ibid., 1979, No.8, pp. 105-108.

54. Korshunov, V. A., Determination of the Phonon Density of States with Thermodynamic Functions of the Crystal: Niobium and Tantalum. - FMM, 1979, 47, issue 6, pp. 1147- 1151.

55. Korshunov, V. A., Tanana, V. P., Determination of the Phonon Density of States with Thermodynamic Functions of the Crystal (Molybdenum, Tungsten, Magnesium, Zinc, Cadmium). - FMM, 1979, **48**, issue 5, pp. 908 - 915.

56. Kotov, B. A., Okuneva, N. M., Plachenova, E. L., Experimental Determination of the Thermal Oscillation Spectrum of the Lattice for White Tin. - FTT, 1968, **10**, issue 2, pp. 513-521.

57. Kotov, B. A., Okuneva, N. M., Shakh - Budagov, A. L., An Experimental Method For Determination of Thermal Oscillations for Coherent Scattering in Crystals. - FTT, 1967, **9**, issue 9, pp. 2556-2558.

58. Kulik, I. O., Omelyanchuk, A. N., Nonequilibrium Fluctuations in Point Contacts of Normal Metals. - FNT, 1984, **10**, No.3, pp. 305-317.

59. Kulik, I. O., Omelyanchuk, A. N., Tuluzov, I. G., Kinetic Inductance of Point Contacts Between Normal Metals. - FNT, 1982, **8**, No.7, pp. 769-773.

60. Kulik, I. O., Omelyanchuk, A. N., Tuluzov, I. G., Sarkisyants, T. Z. - Effects of High Frequency Rectification in Microcontacts Between Normal Metals. - FNT, 1984, **10**, no.8, pp. 882-885.

61. Kulik, I. O., Omelyanchuk, A. N., Tuluzov, I. G., Second Order Processes in Microcontacts of Normal Metals. - FNT, 1984, **10**, no.9, pp. 929 - 940.

62. Kulik, I. O., Omelyanchuk, A. N., Shekhter, R. I. Electrical Conductivity of Point Microcontacts and the Spectroscopy of Phonons and Impurities in Normal Metals. - FNT, 1977, **3**, no.12, pp. 1543 - 1558.

63. Kulik, I. O., Omelyanchuk, A. N., The Josephson Effect in Superconducting Microbridges: Microscopic Theory. - FNT, 1978, **4**, no.3, pp. 296 - 311.

64. Kulik, I. O., Omelyanchuk, A. N., Yanson, I. K., Nonequilibrium Phonons in Point Contacts Between Normal Metals. - FNT, 1981, **7**, no.2, pp. 263 - 267.

65. Kulik, I. O., Thermal Phonon Spectroscopy in Metals With the Aid of Microcontacts - Kharkov, 1984, - 43 pp. - (Preprint / FTINT - Akad. Nauk Ukr. SSR; no. 8 - 84).

66. Kulik, I. O., Shekhter, R. I., Microcontact Spectroscopy of Ferromagnetic Metals. - FNT, 1980, **6**, no. 2, pp. 184-192.

67. Kulik, I. O., Shekhter, R. I., Shkorbatov, A. G., Microcontact Spectroscopy of the Electron-Phonon Interaction in Metals Having a Small Electron Free Path Length. - JETP, 1981, **81**, issue 6, pp. 2126-2141.

68. Kulik, I. O., Yanson, I. K., Microcontact Spectroscopy of Phonons in the Dirty Limit. - FNT, 1978, **4**, no.10, pp. 1267-1278.

69. Lifschitz, I. M., On the Determination of the Energy Spectrum of a Bose System by its Heat Capacity. - JETP, 1954, **26**, issue 5, pp. 551-556.

70. Lysykh, A. A., Khotkevich, A. V., Kamarchuk, G. V., Microcontact Function of the Electron-Phonon Interaction in Thallium. - In the book: XXI International Conference on the Physics and Techniques of Low Temperatures, (Varna, 11-14 Oct. 1983): Collected Papers - Sofia: In-t Physics of the Solid State, In-t electronics. BAN, 1983, pp. 55-57.

71. Lysykh, A. A., Yanson, I. K., Investigation of the Electron-Phonon Interaction Function of Sodium and Lithium by the Method of Microcontact Spectroscopy. -FTT, 1979, vol. 22, issue 1, pp. 117-119.

72. Lysykh, A. A., Yanson, I. K., Shklarevskii, O. I., Naiydyuk, Y. G., Investigation of the Electron-Phonon Interaction in Nickel and Iron by the Method of Microcontact Spectroscopy. -FNT, 1980, **6**, no. 4, pp. 471-478.

73. Lamb, J., Jeklevik, R. K., Molecular Excitations in Barriers. I., - In the book: The Tunnel Effect in the Solid State. M.: - World, 1973, pp. 223-232.

74. Meekhin, N. P., Yanson, I. K., Apparatus for Studies of Small Nonlinearities in the Volt-Amperage Characteristics of Tunneling Contacts: (review). - In the book : Physics of the Condensed State. Kharkov: , FTINT Akad. Nauk Ukr. SSR, 1973, issue 29, pp. 122-131.

75. Mostvoi, V. M., Revenko, Y. F., Svistunov, V. M., The Electron-Phonon Interaction in Alloys of Indium - Magnesium with a Structural Phase Transition. - JETP, 1981, **80**, issue 1, pp. 235-243.

76. Mostvoi, V. M., Revenko, Y. F., The Electron-Phonon Interaction in Hydrostatically Compressed Alloys of Indium and Tin. - FNT, 1981, **7**, no. 5, pp. 658-661.

77. Naidyuk, Yu. G., Chernoplekov, N. A., Shitikov, Y. L., et al., Observation of the Interaction of Electrons with Local Oscillations in Metallic Microcontacts. - JETP, 1982, **83**, issue 3, pp. 1177-1181.

78. Naidyuk, Yu. G., Shklarevskii, O. I., The Electron-Phonon Interaction Spectrum in Beryllium. - FTT, 1982, **24**, issue 9, pp. 2631-2635.

79. Naidyuk, Yu. G., Shklarevskii, O. I., Yanson, I. K., Microcontact Spectroscopy of the Dilute Magnetic Alloys CuMn and CuFe. - FNT, 1982, **8**, no. 7, pp. 725-731.

80. Naidyuk, Yu. G., Yanson, I. K., Lysykh, A. A., Shklarevskii, O. I., A Study of the Electron-Phonon Interaction in Alkaline Metals. - FTT, 1980, **22**, issue 12, pp. 3665-3672.

81. Naidyuk, Yu. G., Yanson, I. K., Lysykh, A. A., Shklarevskii, O. I., The Electron-Phonon Interaction in Microcontacts of Gold and Silver. - FNT, 1982, **8**, no. 9, pp. 922-929.

82. Naidyuk, Yu. G., Yanson, I. K., Lysykh, A. A., Shitikov, Y. L., Microcontact Spectroscopy of α-alloys of Ni-Be. - FTT, 1984, **26**, issue 9, pp. 2734-2738.

83. Naidyuk, Yu. G., Yanson, I. K., Shklarevskii, O. I., The Electron-Phonon Interaction Spectrum in Magnesium. - FNT, 1981, **7**, no. 3, pp. 322-326.

84. Omelyanchuk, A. N., Kulik, I. O., Shekhter, R. I., Towards the Theory of Nonlinear Effects in the Electrical Conductivity of Metallic Microcontacts. - Letter in JETP, 1977, **25**, issue 10, pp. 465-469.

85. Omelyanchuk, A. N., A Theory of Weak Superconducting Contacts and Microcontact Spectroscopy of Metals: Author's abstract of dissertation candidate for Phys. - Math. Science. - Kharkov, 1978 - 17 pp.

86. Omelyanchuk, A. N., Tuluzov, I. G. - Nonlinear Electrical Conductivity of Metallic Microcontacts with an Alternating Electric Field. - FNT, 1983, **9**, no. 3, pp. 284-296.

87. Omelyanchuk, A. N., Tuluzov, I. G., The Electrical Conductivity of Metallic Microcontacts Containing Magnetic Impurities. - FNT, 1980, **6**, no. 10, pp. 1286 -1291.

88. Popilov, L. Y., Zaitsev, L. P., Electropolishing and Electroetching of Metallic Specimens. - M. : Metallurgizdat Publishers, 1963, 410 pp.

89. Revenko, Y. F., D'yachenko, A. I., Svistunov, V. M., Shonaikh, B., The Effect of Pressure on the Electron- Phonon Interaction in Niobium and Tantalum. - FNT, 1980, **6**, no. 10, pp. 1304-1313.

90. Revenko, Y. F., Mostvoi, V. M., Svistunov, V. M., Tunneling Spectroscopy of Alloys of α-, β-phase In-Sn and of Hydrostatically Compressed Indium. - FNT, 1981, **7**, no. 2, pp. 141-153.

91. Ribalchenko, L. F., Yanson, I. K., Bobrov, N. L., Fisun, V. V., Microcontact Spectroscopy of Tantalum, Molybdenum and Tungsten. - FNT, 1981, **7**, No. 2, pp. 169-175.

92. Ribalchenko, L. F., Yanson, I. K., Fisun, V. V., Microcontact Spectra of Vanadium. - FTT, 1980, **22**, issue 7, pp. 2028-2033.

93. Svarka in Mechanical Engineering: (reference book): In 4 - x t., N. A. Olshanskii, ed., M.: Mechanical Engineering, 1978 - T. 1., 502 pp.

94. Svistunov, V. M., Belogolovskii, M. A., D'yachenko, A. I., Single-Particle Tunneling in Metal-Insulator-Metal Structures: (review). 2. Spectroscopy of the Electron-Phonon Interaction. - Metallofiz., 1983, **5**, No. 5, pp. 17-25.

95. Svistunov, V. M., Belogolovskii, M. A., Revenko, Y. F., Chernyak, O. I., Tunneling Spectroscopy of Supercurrent Under High Pressure. - Visnik, Akad. Nauk Ukr. SSR, 1981, No. 2, pp. 8-15.

96. Svistunov, V. M., Belogolovskii, M. A., Chenyak, O. I., et. al., Singularities in the Tunneling Conductance of Normal Contacts - JETP, 1983, **84**, issue 5, pp. 1781-1791.

97. Svistunov, V. M., D'yachenko, A. I., Chernyak, O. I., Application of Dispersion Relations to Analysis of Tunneling Data. - FTT, 1977, **19**, issue 6, pp. 1994-2000.

98. Sirota, N. N., Bulam, I. A., The Phonon Spectrum of α-Fe in Measurements Using Thermal Neutrons. - FTT, 1976, **18**, issue 4, pp. 981-984.

99. Sirota, N. N., Bulam, I. A., Experimental Determination of the Phonon Spectrum of Beryllium by Inelastic Neutron Scattering. DAN, SSSR, 1976, **226**, no. 3, pp. 554-557.

100. Smitlz, K., Metals. - reference book - Metallurgia (Moscow) 1980, 447 pp.

101. Sirikh, G. F., Zhernov, A. P., Zemlianov, M. G., et. al., The Effect of Tantalum and Tungsten Impurities on the Phonon Spectrum of Vanadium. - JETP, 1976, **70**, issue 1, pp. 353-359.

102. Tanana, V. P., Korshunov, V. A., The Principle of Minimal Errors. - DAN SSSR, 1978, **239**, No. 4, pp. 800-803.

103. Thomas, D. E., Klein, J. M., Detection of Tunneling Current Structures With the Aid of Differentiation. - Rev. Sci. Inst., 1963, No. 8, pp. 85-89.

104. Thomas, D. E., Klein, J. M., A Highly Sensitive System for Measurement of the Second Harmonic Voltage. - Rev. Sci. Inst., 1965, No. 9, pp. 13-18.

105. Tulina, N. A., - Microcontact Spectroscopy of Monocrystalline Rhenium. - Metallofiz., 1982, **4**, issue 5, pp. 116-118.

106. Tulina, N. A., - Microcontact Spectroscopy of Re-Os, Re-W Alloys. - FNT, 1983, **9**, No. 5, pp. 499-503.

107. Finkel, V. A., Low Temperature x-ray Diffraction of Metals. - Metallurgia (Moscow), 1971, 256 pp.

108. Khlus, V. A., Nonlinear Voltage-Current Characteristics of a Microcontact of the S-c-N Type. - FNT, 1983, **9**, No. 9, pp. 985-988.

109. Khlus, V. A., Omelyanchuk, A. N., The Electron-Phonon Interaction in Superconducting Microcontacts. - FNT, 1983, **9**, No. 4, pp. 373-384.

110. Holm, R., Electrical Contacts. - M. : Pub. House of Foreign Literature, 1961. - 464 pp.

111. Khotkevich, A. V., Elenskii, V. A., Kovtun, G. P., Yanson, I. K., Microcontact Spectra and Functions of the Electron -Phonon Interaction in Osmium. - FNT, 1984, **10**, No. 4, pp. 375-380.

112. Khotkevich, A. V., Kamarchuk, G. V., Electron-Phonon Interaction Functions in Indium. - In the book: 22-e All Union Conference on the Physics of Low Temperatures (Kishinev, 20 - 30 Oct. 1982): Collected Papers of Kishinev: Institute of Applied Physics, Akad. Nauk MSSR, 1982, chapter 2, pp. 36-37.

113. Khotkevich, A. V., Universal Current Source for Tunneling Research. - Kharkov, 1980 - 7 pp. - Manuscript deposited with VINITI 04-28-80, No. 1716-80.

114. Khotkevich, A. V., Khotkevich, V. V., A. s. (pat. no. 892361 (USSR)), System for Measurement of the Nonlinear Volt - Amperage Characteristics of a Point Contact. - Published in B. I., 1981, No. 47.

115. Khotkevich, A. V., Khotkevich, V. V., A New Method for Measuring Small Nonlinear Volt-Amperage Characteristics of Microscopic Point Contacts. - Kharkov, 1980. - 8 pp. - Manuscript deposited with VINITI 08-11-80, No. 3528 -80.

116. Khotkevich, A. V., Yanson, I. K., A Study of the Electron-phonon Interaction Function in Lead by the Method of Microcontact Spectroscopy. - FNT, 1981, **7**, No. 5, pp. 623-629.

117. Khotkevich, A. V., Yanson, I. K., Microcontact Spectroscopy of the Electron-phonon Interaction in β-Sn., FTT, 1981, **23**, issue 7, pp. 2064-2071.

118. Khotkevich, A. V., Yanson, I. K., Oscillation of the Differential Conductance of a Microscopic Point Contact in Helium. - II - FNT, 1981, **7**, No. 1, pp. 69-81.

119. Khotkevich, A. V., Yanson, I. K., Combined Study of the Energy Dependence of Excess Current in the Superconducting State and the EPI in the Normal State for Point Microcontacts. - FNT, 1981, 7, No. 6, pp. 727-737.

120. Khotkevich, V. V., Khotkevich, A. V., Investigation of the Electron-Phonon Interaction in the Intermetallic Compound Cd_3Mg by the Method of Microcontact Spectroscopy. - FNT, 1985, **11**, No. 5, pp. 491-495.

121. Chernoplekov, N. A., Zemlianov, M. G., Chicherin, A. G., Liashenko, B. G., Investigation of the Phonon Spectrum of Nickel. - JETP, 1963, **44**, issue 3, pp. 858-860.

122. Chubov, P. N., Akimenko, A. I., Yanson, I. K., A. s. (pat. no. 83408 (USSR)). A Method for Obtaining Pressed Microcontacts Between Metallic Electrodes. - Published in V.I., 1981, No. 20.

123. Chubov, P. N., Yanson, I. K., Akimenko, A. I., The Electron -Phonon Interaction in Aluminum Microcontacts. - FNT, 1982, **8**, No. 1, pp. 64-80.

124. Shalov, Y. N., Yanson, I. K., The Electron-Phonon Interaction Spectrum in Silver and Gold. - FNT, 1977, **3**, No. 1, pp. 99-102.

125. Sharvin, Y. V., On One Possible Method for Investigating the Fermi Surface. - JETP, 1965, **48**, issue 3, pp. 984-985.

126. Shekhter, R. I., Kulik, I. O., The Spectroscopy of Phonons in Heterocontacts. - FNT, 1983, **9**, No.1, pp. 46-55.

127. Shekhter, R. I., Spectroscopy of the Energy Dependence of the Energy Relaxation Time of Hot Electrons in Semiconductors using Microcontacts. - FTP, 1983, **17**, No. 8, pp. 1463-1470.

128. Shklarevskii, O. I., Gribov, N. N., Naidyuk, Y. G., The Electron-Phonon Interaction in Microcontacts of Gallium. - FNT, 1983, **9**, No. 10, pp. 1068-1077.

129. Shklarevskii, O. I., Naidyuk, Yu. G., Yanson, I. K., Microcontact Spectroscopy of the Electron-Phonon Interaction in Intermetallic Cu_3Au. - FNT, 1982, **8**, No. 10, pp. 1073-1077.

130. Shustov, G. R., Kulik, I. O., Pseudopotential Calculation of the Electron-Phonon Interaction Function in Metallic Microbridges. - FNT, 1983, **9**, No. 2, pp. 161-169.

131. Adler, J. G., Jackson, J. E., A System for Observing Small Nonlinearities in Tunneling. - Rev. Sci. Inst., 1966, **8**, pp. 63-68.

132. Yanson, I. K., Batrak, A. G., An Investigation of the Anisotropy of the Electron-Phonon Interaction in Zinc by the Method of Microcontact Spectroscopy. - JETP, 1978, **76**, issue 1, pp. 325-339.

133. Yanson, I. K., Batrak, A. G., Microcontact Spectroscopy of the Anisotropy of the Electron-Phonon Interaction in Metals. - Letter in JETP, 1978, **27**, issue 4, pp. 212-216.

134. Yanson, I. K., Bobrov, N. L., Ribalchenko, L. F., Fisun, V. V., Phonon Features of the Excess Current in Superconducting Point Contacts: Nb and Nb_3Sn - In the book: 23 - e All Union Conference on Physics at Low Temperatures (Tallin, 23-25 Oct. 1984): Collected Papers of Tallin: Inst. of Chem. and Biol. Physics - Akad. Nauk ESSR, 1984, ch. 1, pp. 122-123.

135. Yanson, I. K., Bobrov, N. L., Ribalchenko, L. F., Fisun, V. V., Phonon Spectroscopy in Dirty Superconducting Contacts. - FNT, 1983, **9**, No. 11, pp. 1155-1165.

136. Yanson, I. K., Kamarchuk, G. V., Khotkevich, A. V., Nonlinear Voltage-Current Characteristics of Superconductor-Normal Metal Point Contacts, Dependence on the Electron-Phonon Interaction. - FNT, 1984, **10**, No. 4, pp. 415-418.

137. Yanson, I. K., Critical Current and the Voltage-Current Characteristics of Superconducting Microbridges. - FNT, 1975, **1**, No. 2, pp. 141-153.

138. Yanson, I. K., Kulik, I. O., Microcontact Spectroscopy of Phonons in Metals. - Kharkov, 1978. - 25 pp. - (Preprint) FTINT Akad. Nauk Ukr. SSR; No. 24.

139. Yanson, I. K., Microcontact Spectroscopy of the Electron-phonon Interaction in Zinc and Cadmium. - FNT, 1977, **3**, No. 12, pp. 1516-1529.

140. Yanson, I. K., Microcontact Spectroscopy of the Electron-phonon Interaction in Clean Metals: (review article) - FNT, 1983, **9**, No. 7, pp. 676-709.

141. Yanson, I. K., Naidyuk, Yu. G., Shklarevskii, O. I., Two-Phonon Electron Scattering Processes in Metallic Microcontacts. - FNT, 1982, **8**, No. 11, pp. 1178-1186.

142. Yanson, I. K., Nonlinear Effects in the Conductivity of Point Contacts and the Electron-Phonon Interaction in Normal Metals. - JETP, 1974, **66**, issue 3, pp. 1035-1049.

143. Yanson, I. K., The Electron-Phonon Interaction Spectrum in Indium. - FTT, 1974, **16**, issue 12, pp. 3595-3601.

144. Yanson, I. K., Shalov, Y. N., The Electron-Phonon Interaction Spectrum in Copper. - JETP, 1976, **71**, issue 7, pp. 286-299.

145. Akimenko, A. I., Verkin, A. B., Yanson, I. K., Point-contact noise spectroscopy of phonons in metals. - J. Low Temp. Phys., 1984, **54**, No. 3/4, pp. 247 - 266.

146. Allen, P. B., Electron-phonon effects in infrared properties of metals. - Phys. Rev. B, 1971, **3**, No. 2, pp. 305 - 320.

147. Arnold, G. B., Zasadzinski, J., Osmun, J. W., Wolf, E. L., Proximity electron tunneling spectroscopy. II. Effects of the induced N-metal pair potential on calculated S-metal properties. - J. Low temp. Phys. 1980, **40**, No. 3/4, pp. 225 - 246.

148. Ashraf, M., Swihart, J. C., Point contact spectra for sodium and potassium. - Phys. Rev. B, 1982, **25**, No. 4, pp. 2094 - 2130.

149. Ashraf, M., Swihart, J. C., Possible direct observation of phasons in potassium. - Phys. Rev. Lett., 1983, <u>50</u>, No. 12, pp. 921 - 924.

150. Beaulac, T. P., Allen, P. B., Pinski, F. J., Electron-phonon effects in copper. II. Electrical and thermal resistivities and Hall coefficient. - Phys. Rev. B, 1982, **26**, No. 4, pp. 1549 - 1558.

151. Beg, M. M., Nielsen, M., Temperature dependence of lattice dynamics of lithium 7. - Ibid., 1976, **14**, No. 10, pp. 4266 - 4273.

152. Behari, J., Tripathi, B. B., Angular forces and lattice dynamics of fcc metals. - Indian J. Pure Appl. Phys., 1971, **9**, No. 12, pp. 1037 - 1041.

153. Behari, J., Tripathi, B. B., Lattice dynamics of some bcc transition metals. - J. Phys. Soc. Japan, 1972, **33**, No. 5, pp. 1207 - 1213.

154. Bergsma, J., van Disk, C., Tochetti, D., Normal vibration in α-iron. - Phys. Lett., 1967, **24A**, No. 5, pp. 270 - 272.

155. Bhatia, A. B., Horton, G. K., Vibration spectra and specific heats of cubic metals. II. Applications to silver. - Phys. Rev., 1955, **98**, No. 6, pp. 1715 - 1721.

156. Bhatia, A. B., Vibration spectra and specific heats of cubic metals. I. Theory and applications to sodium. - Ibid., **97**, No. 2, pp. 363 - 371.

157. Birgeneau, R. J., Cardes, J., Dolling, G., Woods, A. D. B., Normal modes of vibrations in nickel. - Ibid., 1964, **136**, No. 5A, pp. 1359 - 1365.

158. Blonder, G. E., Tinkham, M., Klapwijk, T. M., Transition from metallic to tunneling regimes in superconducting microconstrictions: Excess current, charge imbalance, and supercurrent conversion. - Phys. Rev. B, 1982, **25**, No. 7, pp. 4515 - 4532.

159. Blonder, G. E., Tinkham, M., Metallic to tunneling transition in Cu-Nb point contacts. - Ibid., 1983, **27**, No. 1, pp. 112 - 118.

160. Bose, G., Tripathi, B. B., Gupta, H. C., The lattice dynamics of hexagonal close packed metals. - J. Phys. Soc. Japan, 1973, **34**, No. 4, pp. 1006 - 1013.

161. Bostock, M. H., Diddiuk, V., Chaung, W. H., et al., Does strong-coupling theory describe superconducting Nb? - Phys. Rev. Lett., 1976, **36**, No. 11, pp. 603 - 605.

162. Burnell, D., Wolf, E. L., Tunneling determination of electron-phonon interaction in Mg. - Phys. Lett., 1982, **90A**, No. 9, pp. 471 - 473.

163. Burrafato, G., Mancini, N. A., Electron-phonon interaction by numerical analysis of superconducting tunneling experimental data. - Nuovo Cim., 1983, **D2**, No. 5, pp. 1383 - 1400.

164. Bussian, B., Frankowski, I., Wohlleben, D., Metallic point contact spectra of valence fluctuation compounds. - Phys. Rev. Lett., 1982, **49**, No. 14, pp. 1026 - 1029.

165. Carbotte, J. P., Dynes, R. C., Superconductivity in simple metals. - Phys. Rev., 1968, **172**, No. 2, pp. 476 - 484.

166. Carbotte, J. P., Dynes, R. C., Trofimenkoff, P. N., Phonon renormalization of the electronic effective mass. - Can J. Phys., 1969, **47**, No. 10, pp. 1107 - 1116.

167. Carbotte, J. P., Truant, P. T., Dynes, R. C., Electron specific heat of aluminum based alloys. - Ibid., 1970, **48**, No. 12, pp. 1504 - 1513.

168. Caro, J., Point-contact spectroscopy on palladium-hydride, palladium-deuteride, and some transition metals. - Amsterdam, Rodopi, 1983 - 174 pp.

169. Caro, J., Coehorn, R., de Groot, D. G., Direct measurement of the electron-phonon interaction in Pd, Mo and W by point- contact spectroscopy. - Solid State Commun., 1981, **39**, No. 2, pp. 267 - 271.

170. Caro, J., de Groot, D. G., Coehorn, R., Griessen, R., Point- contact spectroscopy of electron-phonon interaction in metals: a study of spectral intensities. - In: Superconductivity in d- and f- band metals, W. Buckel and W. Weber, eds., Karlsruhe, 1982, pp. 573 - 578.

171. Carter, R. C., Hughs, D. J., Palevski, H., Inelastic scattering of low-energy neutrons by lattice vibrations of vanadium. - Phys. Rev., 1956, **104**, No. 1, pp. 271 - 272.

172. Chesser, N. J., Axe, J. D., Lattice dynamics of zinc: Phonon structure factor. - Phys. Rev. B, 1974, **9**, No. 10, pp. 4060 - 4067.

173. Clark, B. C., Gazis, D. C., Wallis, R. F., Frequency spectra of body-centered cubic lattices. - Phys. Rev., 1964, **134**, No. 6A, pp. A1487 - A1491.

174. Colella, R. Batterman, B. W., X-ray determination of phonon dispersion in vanadium. - Phys. Rev. B, 1970, **1**, No. 10, pp. 3913 - 3921.

175. Coulthard, M. A., Dynamic electron-phonon coupling functions. - J. Phys. F, 1971, **1**, No. 2, pp. 188 - 194.

176. Cowley, E. R., A Born-von Karman model for lead. - Solid State Commun., 1974, **14**, No. 7, pp. 587 - 589.

177. Cowley, R. A., Woods, A. D. B., Dolling, G., Crystal dynamics of potassium. - Phys. Rev., 1966, **150**, No. 2, pp. 487 - 497.

178. D'Ambrumenil, N., Duif, A. M., Jansen, A. G. M., Wyder, P., Point contact spectroscopy of internal field distributions in spin glasses. - J. Magn. and Magn. Mater., 1983, **31/34**, part 3, pp. 1415 - 1416.

179. D'Ambrumenil, N., White, R. M., Point contact spectroscopy of g-values in metals. - J. Appl. Phys., 1982, **53**, No. 3, pp. 2052 - 2054.

180. Dayan, M., Tunneling measurements on crystalline and granular aluminum. - J. Low Temp. Phys., 1978, **32**, No. 3/4, pp. 643 - 655.

181. Dixon, A. E, Woods, A. D. B., Brockhouse, B. N., Frequency distribution of the lattice vibrations in sodium. - Proc. Phys. Soc., 1963, **81**, No. 523, pp. 973 - 974.

182. Dynes, R. C., Influence of the phonon spectra of In - Tl alloys on the superconducting transition temperatures. - Phys. Rev. B, 1970, **2**, No. 3, pp. 644 - 656.

183. Dynes, R. C., Carbotte, P. T., Preferred direction tunneling into a superconductor. - Physica, 1971, **55**, pp. 462 - 470.

184. Dynes, D. E., Rowell, J. M., Tunneling determination of phonon-energy uncertainties due to force-constant disorders in the alloy $Pb_{1-2x} Tl_x Bi_x$. - Phys. Rev., 1969, **187**, No. 3, pp. 821 - 827.

185. Eisenhauer, C. M., Pelah, I., Hughes, D. J., Palevsky, H., Measurement of lattice vibrations in vanadium by neutron scattering. - Ibid., 1958, **109**, No. 4, pp. 1046 - 1051.

186. Eschrig, H., van Loven, L., Ziesche, P., On the determination of the phonon density of states by means of coherent inelastic scattering of neutrons on polycrystalline samples. - Phys. status solidi (B), 1974, **66**, No. 2, pp. 587 - 593.

187. Flinn, P. A., McManus, G. M., Lattice vibrations and Debye temperatures of aluminum. - Phys. Rev., 1963, **132**, No. 6, pp. 2458 - 2460.

188. Flinn, P. A., McManus, G. M., Rayne, J. A., Effective X-ray calorimetric Debye temperature for copper. - Ibid., 1961, **123**, No. 3, pp. 809 - 812.

189. Franck, J. P., Keeler, W. J., Wu, T. M., Pressure-dependence of the phonon spectrum of Pb from tunneling measurements. - Solid State Commun., 1969, **7**, No. 5, pp. 483 - 486.

190. Frankowski, I., Wachter, P., Point contact spectroscopy of intermediate valence compounds. - J. Appl. Phys., 1982, **53**, No. 11, pp. 7887 - 7889.

191. Frankowski, I. Wachter, P., Point contact spectroscopy on LaS, CdS and TmSe. - Solid State Commun., 1981, **40**, No. 9, pp. 885 - 888.

192. Frankowski, I., Wachter, P., Point contact spectroscopy on SmB_6, TmSe, LaB_6 and LaSe. - Ibid., 1982, **41**, No. 8, pp. 577 - 580.

193. Garret, D. G., Swuhart, J. C., Third-order perturbation theory and exchange and correlation in the lattice dynamics of indium. - J. Phys. F, 1976, **6**, No. 10, pp. 1781 - 1788.

194. Gärtner, K., Hahn, A., Eigenschaften von Tunnelkontakten auf den fünfwertigen supraliteden Übergangsmetallen Nb, Ta und V. - Z. Naturforsch., 1976, **31a**, No. 8, pp. 861 - 871.

195. Geldart, D. J. W., Taylor, R., Varshni, Y. P., Influence of the static electron gas screening function on the lattice dynamics of sodium. - Can. J. Phys., 1970, **48**, No. 2, pp. 183 - 192.

196. Gerlach-Meyer, U., Queisser, H. J., Thermovoltaic evidence for electronic Knudsen flow through silicon microcontacts. - Phys. Rev. Lett., 1983, **51**, No. 20, pp. 1904 - 1906.

197. Gilat, G., Existence of an infinity in the frequency distribution $g(v)$ of monatomic body centered cubic crystals. - Phys. Rev., 1967, **157**, No. 3, pp. 540 - 543.

198. Gilat, G., Phonon density of states in lead. - Solid State Commun., 1965, **3**, No. 5, pp. 101 - 103.

199. Gilat, G., Nicklow, R. M., Normal vibrations in aluminum and derived thermodynamic properties. - Phys. Rev., 1966, **143**, No. 2, pp. 487 - 494.

200. Gilat, G., Raubenheimer, L. J., Accurate numerical methods for calculating frequency-distribution functions in solids. - Ibid., **144**, No. 2, pp. 390 - 395.

201. Grimvall, G., The electron-phonon interaction in metals. - Amsterdam, North Holland, 1981. - 304 pp.

202. Grimvall, G., The electron-phonon interaction in normal metals. - Phys. Scripta, 1976, **14**, No. 1/2, pp. 63 - 78.

203. Gupta, H. C., Tripathi, B. B., Lattice dynamics of lead: an elastic force model approach. - Phys. status solidi (b), 1971, **45**, No. 1, pp. 235 - 239.

204. Gupta, O. P., Crystal dynamics of face centered cubic cobalt. - Solid State Commun., 1982, **42**, No. 1, pp. 31 - 32.

205. Gupta, R. P., Vibration spectra and Debye temperatures of silver and gold. - J. Phys. Soc. Japan, 1966, **21**, No. 9, pp. 1726 - 1729.

206. Gupta, R. P., Vibration spectrum and specific heat of lithium. - J. Chem. Phys., 1966, **45**, No. 11, pp. 4019 - 4022.

207. Hartmann, W. M., Milbrodt, T. O., Model-potential calculations of phonon energies in aluminum. - Phys. Rev. B, 1971, **3**, No. 12, pp. 4133 - 4143.

208. Hayman, B., Carbotte, J. P., A model for the electron-phonon interaction in polyvalent metals: Al. - Phys. status solidi (b), 1974, **65**, No. 2, pp. 439 - 448.

209. Hayman, B., Carbotte, J. P., Hall coefficient in the alkali metals. - Phys. Rev. B, 1972, **6**, No. 4, pp. 1154 - 1162.

210. Hayman, B., Carbotte, J. P., Resistivity and electron-phonon interaction in Li. - J. Phys. F, 1971, **1**, No. 6, pp. 828 - 835.

211. Hayman, B., Carbotte, J. P., Transport in some alkalies. - Can. J. Phys., 1971, **49**, No. 14, pp. 1952 - 1966.

212. Hendricks, J. B., Riser, H. N., Clark, C. B., Calculated vibrational spectra and specific heats of lithium and vanadium. - Phys. Rev., 1963, **130**, No. 4, pp. 1377 - 1380.

213. Houmann, J. C. G., Nicklow, R. M., Lattice dynamics of terbium. - Phys. Rev. B, 1970, **1**, No. 10, pp. 3943 - 3952.

214. Igalson, J., Pindor, A. J., Sniadower, L., Electron-phonon interaction function for metals from temperature dependence of the electrical resistivity. - J. Phys. F, 1981, **11**, No. 5, pp. 995 - 1010.

215. Itskovich, I. F., Kulik, I. O., Shekhter, R. I., Point-contact spectroscopy of electron-phonon interaction in semiconductors. - Solid State Commun., 1984, **50**, No. 5, pp. 421 - 424.

216. Jacobsen, E. H., Elastic spectrum of copper from temperature-diffuse scattering of X-rays. - Phys. Rev., 1955, **97**, No. 3, pp. 654 - 659.

217. Jansen, A. G. M., Mueller, F. M., Wyder, P., Direct measurement of $\alpha^2 F$ in normal metals using point contacts: noble metals. - In: Superconductivity in d- and f-band metals., D. H. Douglas, ed., - New York, Plenum, 1976, pp. 607 - 623.

218. Jansen, A. G. M., Mueller, F. M., Wyder, P., Direct measurement of electron-phonon coupling $\alpha^2 F$ using point contacts: noble metals. - Phys. Rev. B, 1977, **16**, No. 4, pp. 1325 - 1328.

219. Jansen, A. G. M., Mueller, F. M., Wyder, P., Normal metallic point contacts. - Science, 1978, **199**, No. 4333, pp. 1037 - 1040.

220. Jansen, A. G. M., van den Bosch, J. H., van Kempen, H., et al., Point-contact spectroscopy in alkali metals: K, Na and Li. - J. Phys. F, 1980, **10**, No. 2, pp. 265 - 273.

221. Jansen, A. G. M., van Gelder, A. P., Wyder, P., Point-contact spectroscopy in metals. - J. Phys. C, 1980, **13**, No. 33, p. 6073 - 6118.

222. Jansen, A. G. M., van Gelder, A. P., Wyder, P., Strässler, S., Application of point-contact spectroscopy in metals to the Kondo problem. - J. Phys. F, 1981, **11**, No. 1, pp. L15 - L21.

223. Joshi, S. K., Hemkar, M. P., Vibrational spectrum and specific heat of potassium. - Physica, 1961, **27**, No. 8, pp. 793 - 796.

224. Joshi, S. K., Hemkar, M. P., Vibrational spectrum of sodium. - Phys. Rev., 1962, **126**, No. 5, pp. 1687 - 1690.

225. Kahn, F. S., Allen, P. B., Butler, W. H., Pinski, F. J., Electron-phonon effects in copper. - Phys. Rev. B, 1982, **26**, No. 4, pp. 1538 - 1548.

226. Kam, Z., Gilat, G., Accurate numerical method for calculating frequency distribution in solids. III. Extension to tetragonal crystals. - Phys. Rev., 1968, **175**, No. 3, pp. 1156 - 1163.

227. Kam, C., Gilat, G., Optimum-model-potential lattice dynamics of β-Sn. - Phys, Rev. B, 1972, **5**, No. 8, pp. 2887 - 2896.

228. Kamal, M., Malik, S. S., Neutron incoherent elastic scattering study of the temperature dependence of the Debye-Waller exponent in vanadium. - Ibid., 1978, **18**, No. 4, pp. 1609 - 1617.

229. Kharoo, H. L., Gupta, O. P., Hemkar, M. P., Angular forces and normal vibration in nickel. - Ibid., 1979, **19**, No. 6, pp. 2986 - 2991.

230. Kharoo, H. L., Gupta, O. P., Hemkar, M. P., Treatment of bcc transition metals on a modified CGW model. - Czech. J. Phys., 1978, **B28**, No. 1, pp. 77 - 83.

231. Klein, J., Leger, A., Belin, M., et al., Inelastic electron tunneling spectroscopy of metal-insulator-metal junctions. - Phys. Rev. B, 1973, **7**, No. 6, pp. 2336 - 2348.

232. Knorr, K., Barth, N., Electron tunneling into disordered thin films. - J. Low Temp. Phys., 1971, **4**, No. 3, pp. 469 - 484.

233. Knorr, K., Barth, N., Superconductivity and phonon spectra of disordered thin films. - Solid State Commun., 1970, **8**, No. 13, pp. 1085 - 1088.

234. Kochler, W. C., Child, H. R., Nicklow, R. M., et al., Spin- wave dispersion relations in gadolinium. - Phys. Rev. Lett., 1970, **24**, No. 1, pp. 16 - 18.

235. Krebs, K., Dispersion curves and lattice frequency distribution of metals. - Phys. Rev., 1965, **138**, No. 1A, pp. A143 - A147.

236. Kulik, I. O., Shekhter, R. I., Omelyanchuk, A. N., Electron-phonon coupling and phonon generation in normal metal microbridges. - Solid State Commun., 1977, **23**, No. 5, pp. 301 - 303.

237. Kulik, I. O., Shekhter, R. I., Point-contact spectroscopy of electron relaxation mechanisms in semiconductors. - Phys. Lett., 1983, **98A**, No. 3, pp. 132 - 134.

238. Leavens, C. R., Carbotte, J. P., Gap anisotropy in weak coupling superconductor. - Ann. Phys., 1972, **70**, No. 2, pp. 338 - 377.

239. Leppin, H. P., Wohlleben, D. K., Point-contact spectroscopy of magnons in metals. - J. Less Common Metals, 1978, **62**, No. 2, pp. 303 - 318.

240. Leslie, J. D., Chen, J. T., Chen, T. T., An electron tunneling investigation of quench-condensed superconductors. - Can. J. Phys., 1970, **48**, No. 23, pp. 2783 - 2803.

241. Leung, H. K., Carbotte, J. P., Taylor, D. W., Leavens, C. R., Multiple plane wave calculation of the electron- phonon interaction in Al. - Ibid., 1976, **54**, No. 15, pp. 1585 - 1599.

242. Lynn, J. W., Smith, H. G., Nicklow, R. M., Lattice dynamics of gold. - Phys, rev. B, 1973, **8**, No. 8, pp. 3493 - 3499.

243. Lysykh, A. A., Yanson, I. K., Shklyarevski, O. I., Naidyuk, Yu. G., Point-contact spectroscopy of electron-phonon interaction in alloys. - Solid State Commun., 1980, **35**, No. 12, pp. 987 - 989.

244. MacDonald, A. H., leavens, C. R., Calculated point-contact electron-phonon spectral functions for the alkali metals. - Phys. Rev. B, 1982, **26**, No. 8, pp. 4293 - 4298.

245. MacDonald, A. H., Leavens, C. R., Contact shape and the strength of the electron-phonon interaction determined by point contact spectroscopy. - Solid State Commun., 1984, **50**, No. 5, pp. 467 - 468.

246. Macdonald, A. H., Leavens, C. R., Influence of elastic scattering on the current-voltage characteristics of small metallic contacts: I. The ohmic current. - J. Physics F, 1983, **13**, No. 3, pp. 665 - 673.

247. MacDonald, A. H., Leavens, C. R., Influence of elastic scattering on the current-voltage characteristics of small metallic contacts: II. Point contact spectroscopy. - J. Phys. F, 1984, **14**, No. 4, pp. 963 - 971.

248. MacDonald, A. H., Leavens, C. R., Influence of electron-electron scattering on point-contact characteristics in simple metals. - Ibid., 1982, **12**, No. 10, pp. 2323 - 2329.

249. Mahesh, P. S., Dayal, B., Lattice dynamics and specific heats of some transition metals on Krebs's model. - Phys. Rev., 1966, **143**, No. 2, pp. 443 - 451.

250. Maxwell, J. C., A treatise on electricity and magnetism. - Oxford, Clarendon, 1904.

251. McLean, A. B., Lonzarich, G. G., Microcontact spectroscopy in praseodymium. - J. Phys. F, 1984, **14**, No. 9, pp. L185 - L190.

252. McMillan, W. L., Rowell, J. M., Lead phonon spectrum calculated from superconducting density of states. - Phys. Rev. Lett., 1965, **14**, No. 4, pp. 108 - 112.

253. McMillan, W.L., Rowell, J.M., Tunneling and strong-coupling superconductivity. - In: Superconductivity, R. D. Parks, ed., New York, Dekker, 1969, vol. 1, pp. 561 - 614.

254. Metzbower, E. A., Noncentral force model for hexagonal close packed crystal lattices. - Phys. status solidi, 1967, **20**, No. 2, pp. 681 - 692.

255. Miller, A. P., Brockhouse, B. N., Crystal dynamics and electronic specific heats of palladium and copper. - Can. J. Phys., 1971, **49**, No. 6, pp. 704 - 723.

256. Minkiewicz, V. J., Shirane, G., Nathans, R., Phonon dispersion relations for iron. - Phys. Rev., 1967, **162**, No. 3, pp. 528 - 531.

257. Moller, H. B., Houmann, J. C. C., Mackintosh, A. P., Magnetic interaction in Tb and Tb - 10% Ho from neutron scattering. - J. Appl. Phys., 1968, **39**, No. 2, pp. 807 - 815.

258. Musgrave, M. J. P., On the relation between grey and white tin (α-Sn and β-Sn). - Proc. Roy. Soc. A, 1963, **272**, No. 1351, p. 503 - 528.

259. Nakagawa, Y., Woods, A. D. B., Lattice dynamics of niobium. - Phys. Rev. Lett., 1963, **11**, No. 6, pp. 271 - 276.

260. Nicklow, R. M., Gilat, G., Smith, H. G., et al., Phonon frequencies in copper at 49 and 298 K. - Phys. Rev., 1967, **164**, No. 3, pp. 922 - 928.

261. Nicklow, R. M., Wakabayachi, N., Vijayaradhavan, P. R., Lattice dynamics of holmium. - Phys. Rev. B, 1971, **3**, No. 4, pp. 1229 - 1234.

262. Nilsson, G., Rolandson, S., Lattice dynamics of copper at 80 K. - Ibid., 1973, **7**, No. 6, pp. 2393 - 2400.

263. Octavio, M., Tinkham, M., Blonder, G. E., Klapwijk, T. M., Subharmonic energy-gap structure in superconducting constrictions. - Ibid., 1983, **27**, No. 11, pp. 6739 - 6746.

264. Pădureanu, I., Răpeanu, S., Crăcium, C., Phonon density of states in aluminium and bismuth. - Rev. Roum. Phys., 1979, **24**, No. 5, pp. 501 - 507.

265. Page, D. I., The phonon frequency distribution of vanadium. - Proc. Phys. Soc., 1967, **91**, No. 1, pp. 76 - 85.

266. Pal, S., Lattice dynamical properties of nickel. - Can. J. Phys., 1973, **51**, No. 17, pp. 1869 - 1873.

267. Pal, S., Lattice vibrations in palladium. - J. Phys. F, 1971, **1**, No. 5, pp. 588 - 592.

268. Pal, S., Lattice vibrations in vanadium. - Austral. J. Phys., 1974, **27**, No. 4, pp. 471 - 479.

269. Pal, S., Singh, R. B., Phonon dispersion, frequency spectrum and specific heat of palladium. - J. Phys. Soc. Japan, 1973, **35**, No. 5, pp. 1487 - 1491.

270. Pathak, L. P., Rai, R. C., Hemkar, M. P., Lattice vibrations and Debye temperatures of transition metals. - J. Phys. Soc. Japan, 1978, **44**, No. 6, pp. 1834 - 1838.

271. Pepper, M., Ballistic injection of electrons in metal- semiconductor junctions. I. Phonon spectroscopy and impurity enhanced inelastic scattering in n^+ silicon. - J. Phys. C., 1980, **13**, No. 26, pp. L709 - L716.

272. Pepper, M., Ballistic injection of electrons in metal- semiconductor junctions. II. Phonon spectroscopy of aluminium. - J. Phys. C., 1980, **13**, No. 26, pp. L717 - L719.

273. Pepper, M., Ballistic injection of electrons in metal- semiconductor junctions. III. Phonon and sub band spectroscopy of silicon inversion layers. - J. Phys. C., 1980, **13**, No. 26, pp. L721 - L723.

274. Phillips, J. C., Critical points and lattice vibration spectra. - Phys. Rev., 1956, **104**, No. 5, pp. 1263 - 1277.

275. Pindor, A. J., A new method of obtaining information about phonon density of states and electron-phonon interaction in metals. - Warzawa, 1980. - 12p. - (Preprint/Inst. phys. Pol. acad. sci.).

276. Pinski, F. J., Allen P. B., Butler, W. H., Calculated electron-phonon coupling and superconducting T_C of transition metals: Mo and Pd. - J. de Phys., 1978, **39**, coll. C6, pp. 472 - 473.

277. Powell, B. M., Martel, P., Woods, A. D. B., Phonon properties of niobium, molybdenum and their alloys. - Can. J. Phys., 1977, **55**, No. 18, pp. 1601 - 1612.

278. Prakash, J., Semwal, B. S., Sharma, P. K., Phonon frequency distribution of copper, nickel, and vanadium. - Acta phys. Hung., 1971, **30**, No. 3, pp. 231 - 240.

279. Pynn, R., Squires, G. L., Measurements of normal frequencies of magnesium. - Proc. Roy. Soc. A, 1972, **326**, No. 1566, p. 347 - 360.

280. Ramamurthy, V., Rajendraprasad, S. B., Lattice dynamical study of face centered tetragonal indium. - Can. J. Phys., 1983, **61**, No. 1, pp. 58 - 66.

281. Rao, R. R., Menon, C. S., Lattice dynamics, third order elastic constants and thermal expansion of gadolinium. - J. Phys. Chem. Solids, 1974, **35**, No. 3, pp. 425 - 432.

282. Rao, R. R., Ramanand, A., Lattice dynamics, third order elastic constants and thermal expansion of rhenium. - Ibid., 1977, **38**, No. 8, pp. 831 - 835.

283. Rao, R. R., Ramanand, A., Lattice specific heats of cobalt and ruthenium. - Phys. Rev. B, 1979, **19**, No. 4, pp. 1972 - 1975.

284. Raubenheimer, L. J., Gilat, G., Accurate numerical method of calculation of the frequency distribution function in solids. II. Extension to hcp crystals. - Phys. Rev., 1967, **157**, No. 3, pp. 586 - 599.

285. Reissland, J. A., Ese, O., The lattice dynamics of metals using the Shaw model potential. - J. Phys. F, 1975, **5**, No. 1, pp. 110 - 120.

286. Robinson, B., Geballe, T. H., Rowell, J. M., Tunneling study of niobium using aluminum-aluminum oxide-niobium junctions. - In: Superconductivity in d- and f-band metals, D. H. Douglas, ed., New York, Plenum, 1976, pp. 381 - 386.

287. Rowe, J. M., Vegelatos N., Rush, J. J., Acoustic model of the phonon dispersion relation of NbD_x alloys. - Phys. Rev. B, 1975, **12**, No. 8, pp. 2959 - 2964.

288. Rowell, J. M., Dynes, R. C., Comparison of phonon spectra determined from tunneling experiments and neutron scattering. - In: Phonons: Proc. Intern. conf. Rennes, France, 1971, M. A. Nusimovici, ed., Paris, Flammarion, 1971, pp. 150 - 154.

289. Rowell, J. M., McMillan, W. L., Feldmann, W. L., Phonon spectra in Pb and $Pb_{40}Tl_{60}$ determined by tunneling and neutron scattering. - Phys. Rev., 1969, **178**, No. 3, pp. 897 - 899.

290. Rowell, J. M., McMillan, W. L., Feldmann, W. L., Superconductivity and lattice dynamics of white tin. - Phys. Rev. B, 1971, **3**, No. 12, pp. 4062 - 4073.

291. Roy, A. P., Brockhouse, B. N., Lattice frequency spectra of Pb and $Pb_{40}Tl_{60}$ by neutron spectrometry. - Can. J. Phys., 1970, **48**, No. 15, pp. 1781 - 1787.

292. Sauer, H., Keck, K., Non-equilibrium spectroscopy on Al-Al point contacts. - In: Proc. 17th Intern. Conf. on Low Temp. Phys.: LT-17 (Karlsruhe, 15 - 22 Aug. 1984), U. Eckern, et al., eds., Amsterdam, North-Holland, 1984, part 2, pp. 1081 - 1082.

293. Scalapino, D. Y., The electron-phonon interaction and strong coupling superconductors. - In: Superconductivity, R. D. Parks, ed., New York, Dekker, 1969, Vol. 1, pp. 449 - 560.

294. Schöneich, B., Elefani, D., Otschik, P., Schumann, J., Tunnel measurements of Nb single crystals. - Phys. status solidi B, 1979, **91**, No. 1, pp. 99 - 108.

295. Shapiro, S. M., Moss, S. C., Lattice dynamics of face centered cubic $Co_{0.92}Fe_{0.08}$. - Phys. Rev. B, 1977, **15**, No. 3, pp. 2726 - 2730.

296. Sharan, B., Bajpai, R. P., Kumar, A., Lattice dynamical study of cobalt. - Indian J. Pure Appl. Phys., 1970, **8**, No. 3, pp. 135 - 138.

297. Sharan, B., Phonon-frequency distribution curve of vanadium on deLaunay's model. - J. Chem. Phys., 1962, **36**, No. 4, pp. 1116 - 1117.

298. Sharma, P. K., Awasthi, R. K., Noncentral force model for the lattice dynamics of cubic metals. - Phys. Rev. B, 1979, **19**, No. 4, pp. 1963 - 1971.

299. Sharma, P. K., Joshi, S. K., Model for the lattice dynamics of metals. II. Application to face centered cubic metal copper. - J. Chem. Phys., 1964, **40**, No. 3, pp. 662 - 666.

300. Sharma, P. K., Gupta, R. P., Lattice frequency distribution function of vanadium, nickel and aluminium. - Z. Phys. Chem., 1969, **242**, No. 5/6, pp. 341 - 352.

301. Sharp, R. I., The lattice dynamics of niobium. - J. Phys. C, 1969, **2**, No. 3, pp. 421 - 431.

302. Shen, L. Y. Z., Evidence for the electron-phonon interaction in the superconductivity of a transition metal - tantalum. - Phys. Rev. Lett., 1970, **24**, No. 20, pp. 1104 - 1107.

303. Singh, D. N., Bowers, W. A., Vibrational spectrum of vanadium. - Phys. Rev., 1959, **116**, No. 2, pp. 279 - 280.

304. Sinha, S. K., Lattice dynamics of copper. - Ibid., 1966, **143**, No. 2, pp. 422 - 433.

305. Squires, G. L., Relation between the vibration frequencies of a crystal and the scattering of slow neutrons. - Ibid., 1956, **103**, No. 2, pp. 304 - 312.

306. Srivastava, P. L., Dayal, B., Phonon spectrum and specific heats of copper using Bailyn's effective matrix element. - Ibid., 1965, **140**, No. 3A, pp. A1014 - A1019.

307. Stedman, R., Almqvist, L., Nilsson, G., Phonon-frequency distribution and heat capacities of aluminum and lead. - Phys. Rev., 1967, **162**, No. 3, pp. 549 - 557.

308. Stringfellow, M. W., Holdon, T. M., Powell, B. M., Woods, A. D. B., Spin-waves in holmium. - J. Phys. C, suppl., 1970, **3**, No. 2, pp. S189 - S200.

309. Svensson, E. C., Brockhouse, B. N., Rowe, J. M., Crystal dynamics of copper. - Phys. Rev., 1967, **155**, No. 3, pp. 619 - 632.

310. Svensson, E. S., Powell, B. M., Woods, A. D. B., Teuchert, W. D., Phonon dispersion in $Co_{0.92}Fe_{0.08}$. - Can. J. Phys., 1979, **57**, No. 2, pp. 253 - 262.

311. Svistunov, V. M., Chernyak, O. I., Belogovskii, M. A., D'yachenko, A. I., Electron-phonon interaction in lead and lead-indium alloys under pressure. - Phil. Mag. B, 1981, **43**, No. 1, pp. 75 - 92.

312. Swarties, H. M., Jansen, A. G. M., Wyder, P., The Shubnikov- de Haas effect in metallic point contacts. In : Proc. 17th Intern. Conf. on Low Temp. Phys.: LT-17 (Karlsruhe, 15 - 22 Aug., 1984), U. Eckern, et al., eds., Amsterdam, North Holland, 1984, pt. 2, pp. 1083 - 1084.

313. Tevari, V. K., Bullough, R., On a semi-continuum Green's function method for lattice dynamics with application to copper. - J. Phys. F, 1971, **1**, No. 5, pp. 554 - 569.

314. Tomlinson, P. G., Swihart, J. C., Calculation of the cyclotron mass and superconducting energy gap as a function of Fermi surface position in zinc. - Phys. Rev. B, 1979, **19**, No. 4, pp. 1867 - 1891.

315. Tomlinson, P. G., Theory of transport properties of pure single crystal zinc. - Ibid., pp. 1893 - 1904.

316. Trott, A. J., Heald, P. T., The lattice dynamics of the hexagonal close packed metals. - Phys. status solidi (b), 1971, **46**, No. 1, pp. 361 - 368.

317. Truant, P. T., Carbotte, J. P., Electron-phonon function $\alpha^2(\omega)F(\omega)$ for thallium. - Phys. Rev. B, 1972, **6**, No. 10, pp. 3642 - 3648.

318. Truant, P. T., Carbotte, J. P., Theory of phonon induced anisotropy in hcp metals: Zn. - Can. J. Phys., 1973, **51**, No. 9, pp. 922 - 937.

319. Tulina, N. A., Point-contact spectroscopy of rhenium single crystal. - Solid State Commun., 1982, **41**, No. 4, pp. 313 - 315.

320. Van der Heijden, R. W., Jansen, A. G. M., Stoefinga, J. H. M., et al., A new mechanism for high-frequency rectification at low temperatures in point contacts between identical metals. - Appl. Phys. Lett., 1980, **37**, No. 2, pp. 245 - 248.

321. Van Gelder, A. P., A theory for the α^2F dependence of the electrical impedance of a point contact between metals. - Solid State Commun., 1978, **25**, No. 12, pp. 1097 - 1100.

322. Van Gelder, A. P., Jansen, A. G. M., Wyder, P., Temperature dependence of point-contact spectroscopy in copper. - Phys. Rev. B, 1980, **22**, No. 4, pp. 1515 - 1521.

323. Van Gelder, A. P., On the structure of the d^2I/dV^2 characteristics of point contacts between metals. - Solid State Commun., 1980, **35**, No. 1, pp. 19 - 21.

324. Varshni, Y. P., Yuen, P. S., Angular forces in the lattice dynamics of face centered cubic metals. II. - Phys. Rev., 1968, **174**, No. 3, pp. 766 - 769.

325. Vengurlekar, A. S., Inkson, J. C., On the ballistic injection into semiconductors from point contacts and the structure in d^2I/dV^2 characteristics. - Solid State Commun., 1983, **45**, No. 1, pp. 17 - 21.

326. Verkin, B. I., Yanson, I. K., Kulik, I. O., et al., Singularities in d^2V/dI^2 dependences of point contacts between ferromagnetic metals. - Ibid., 1979, **30**, No. 4, pp. 215 - 218.

327. Walker, C. B., X-ray study of lattice vibration frequencies of a crystal and the scattering of slow neutrons. - Phys. Rev., 1956, **103**, No. 3, pp. 547 - 557.

328. Wexler, G., The size effect and the non-local Boltzmann transport equation in orifice and disk geometry. - Proc. Phys. Soc., 1966, **89**, No. 4, pp. 927 - 941.

329. Wolf, E. L., Burnell, D. M., Khim, Z. G., Noer, R. J., Electron-phonon coupling and superconductivity of tantalum. - J. Low Temp. Phys., 1981, **44**, No. 1/2, pp. 89 - 118.

330. Wolf, E. L., Noer, R. J., Arnold, G. B., Proximity electron tunneling spectroscopy. III. Electron-phonon coupling in the Nb-Zr system. - J. Low Temp. Phys., 1980, **40**, No. 5/6, pp. 419 - 440.

331. Wolf, E. L., Zasadzinski, J., Osmun, J. W., Arnold, G. B., Quantitative proximity tunneling spectroscopy. - Solid State Commun., 1979, **31**, No. 5, p. 321 - 324.

332. Wolf, E. L., Zasadzinski, J., Osmun, J. W., Proximity electron tunneling spectroscopy. I. Experiments on Nb. - J. Low Temp. Phys., 1980, **40**, No. 1/2, pp. 19 - 50.

333. Woods, A. D. B., Lattice dynamics of tantalum. - Phys. Rev., 1964, **136**, No. 3A, pp. A781 - A 783.

334. Worlton, T. G., Schmunk, R. E., Lattice vibrations of thallium at 77 and 296 K. - Phys. Rev. B, 1971, **3**, No. 12, pp. 4115 - 4123.

335. Yanson, i. K., Akimenko, A. I., Verkin, A. B., Electrical fluctuations in normal metal point-contacts. - Solid State Commun., 1982, **43**, No. 10, pp. 765 - 768.

336. Yanson, I. K., Bobrov, N. L., Rybalchenko, L. F., Fisun, V. V., Phonon singularities on volt-ampere curves of niobium point contacts. - Ibid., 1984, **50**, No. 6, pp. 515 - 519.

337. Yanson, I. K., Kulik, I. O., Batrak, A. G., Point-contact spectroscopy of electron-phonon interaction in normal metal single crystals. - J. Low Temp. Phys., 1981, **42**, No. 5/6, pp. 527 - 556.

338. Yanson, I. K., Kulik, I. O., Point-contact spectroscopy of phonons in metals. - J. de Phys., 1978, **39**, coll. C6, pp. 1564 - 1566.

339. Young, J. A., Koppel, J. U., Lattice vibrational spectra of beryllium, magnesium and zinc. - Phys. Rev., 1964, **134**, No. 6A, pp. A1476 - A1479.

340. Zasadzinski, J., Burnell, D. M., Wolf, E. L., Arnold, G. B., Superconducting tunneling study of vanadium. - Phys. rev. B, 1982, **25**, No. 3, pp. 1622 - 1632.

Subject Index

Aperture model 1

    - with full or partial reabsorption
      of nonequilibrium phonons 2

Background 27

Background parameter 27

Channel model 1

Contact of two dirty metals 21

    - clean and dirty metals 22

Differential resistance of a point contact 34

Effective volume of phonon generation 10

EPI point contact function 26

Film structures containing point contacts 29

Form coefficient 6

Heterocontact 18

    - clean 20

Point contact 1

    - needle-plane 30

    - sliding type 31

Models of point contacts 1, 2

Modulation broadening function 36

Multiphonon processes 23

Phonon density of states 24

Point contact spectra 10

Quality Criteria 31

Regimes of Electron Transport

    - ballistic 2

    - diffusion 3

    - intermediate 7

Static resistance 36

Systems with concentrated current 1

T model 2

Temperature broadening function 10

Thermodynamic EPI function (Eliashberg function) 24

Transport EPI function 25